Perceptions and Analysis of Digital Risks

Mediating Knowledge Set

coordinated by
Anne Lehmans and Vincent Liquète

Volume 1

Perceptions and Analysis of Digital Risks

Edited by
Camille Capelle
Vincent Liquète

iSTE

WILEY

First published 2021 in Great Britain and the United States by ISTE Ltd and John Wiley & Sons, Inc.

ISTE Ltd
27-37 St George's Road
London SW19 4EU
UK

www.iste.co.uk

John Wiley & Sons, Inc.
111 River Street
Hoboken, NJ 07030
USA

www.wiley.com

Library of Congress Control Number: 2021946232

British Library Cataloguing-in-Publication Data
A CIP record for this book is available from the British Library
ISBN 978-1-78630-744-6

MIX
Paper from responsible sources
FSC® C013604

Contents

Foreword xi
Franc MORANDI

Introduction xvii
Camille CAPELLE

Part 1. Risk Perceptions, Education and Learning 1

Chapter 1. Digital Risks: An Obstacle or a Lever for Education? 3
Camille CAPELLE

1.1. Introduction. 3
1.2. Digital risks and education: what are we talking about? 4
1.2.1. Digital risks 4
1.2.2. What are the risks in education? 8
1.3. Questioning perceptions of digital risks among new teachers 9
1.3.1. Why was this target audience chosen? 9
1.3.2. Methodology and data collection. 10
1.4. Teachers' perceptions of digital risks 11
1.4.1. When perceptions of risk inhibit any practice 11
1.4.2. When perceptions of risk freeze practices 14
1.4.3. When risk perceptions lead us to consider them in order to overcome them 18
1.5. Reflection on the role of digital risk representations in education. . . . 21
1.6. Conclusion 24
1.7. References 25

Chapter 2. Teenagers Faced with "Fake News": Perceptions and the Evaluation of an Epistemic Risk 27

Gilles SAHUT and Sylvie FRANCISCO

2.1. Introduction. 27
2.2. Fake news: From production to reception 28
2.2.1. Characterizing the fake news phenomenon. 29
2.2.2. The potential risks associated with fake news 31
2.2.3. The credibility of fake news 32
2.3. Methodological framework of the study. 34
2.4. Results of the study 36
2.4.1. A heterogeneous understanding of the concept 37
2.4.2. A blurred perception of the goals of fake news 39
2.4.3. The diversity of fake news sources 40
2.4.4. Identifying fake news: heuristic processing and analytical strategies. 42
2.4.5. A remote and controlled phenomenon?. 45
2.5. Discussion of the results and reflections on media and information literacy. 46
2.6. Conclusion 49
2.7. References 50

Chapter 3. "A Big Nebula that is a Bit Scary" (Louise, Trainee Schoolteacher): Training through/in Digital Technology, in School and in Professional Training 55

Anne CORDIER

3.1. Social beings, above all else 57
3.1.1. A "fluid identity" to be grasped 57
3.1.2. Digital technology in the actors' personal ecosystem. 61
3.2. Understanding of digital technology in the classroom. 62
3.2.1. Crystallization and awareness of issues 62
3.2.2. When the socio-technical framework hinders the entry of digital technology into the classroom 64
3.2.3. Rather modest and low-risk experiments 66
3.3. Teaching with and through digital technology: Constant risks. 68
3.3.1. Tensions in the classroom. 68
3.3.2. Tensions in training 71
3.3.3. Desires on both sides. 73
3.4. Potential courses of action 76
3.5. References 78

Part 2. Risks in the Light of Socio-Economic Issues 81

Chapter 4. Top Managers Confronted with Information Risks: An Exploratory Study within the Telecommunications Sector 83

Dijana LEKIC, Anna LEZON-RIVIÈRE and Madjid IHADJADENE

4.1. Introduction. 83
4.2. Information risk: The conceptual field 84
4.3. Controlling information risks: Security policy 89
4.4. Information risk and management 91
4.5. Study methodology and the stakeholder group 93
4.6. Information risk: The perspective of top telecoms managers. 94
4.6.1. Top managers as responsible for information risk management . . 94
4.6.2. Information risk management. 97
4.6.3. Operational challenges related to the information risk management approach 100
4.7. Conclusion 104
4.8 Acknowledgments 106
4.9. References 106

Chapter 5. Cell Phones and Scamming Risks in Cameroon: Users' Experiences and Socio-Institutional Responses 111

Freddy TSOPFACK FOFACK and Abdel Bernazi RENGOU

5.1. Introduction. 111
5.2. Mechanisms behind cell phone scamming in Cameroon: Exhibiting credulity. 115
5.2.1. Setting the scene 116
5.2.2. Enticing but misleading proposals 117
5.2.3. Disguised telephone number confusion 119
5.3. The dynamics of cell phone use in Cameroon 121
5.3.1. The Ministry of Posts and Telecommunications. 121
5.3.2. Agence Nationale des Technologies de l'Information et de la Communication. 122
5.3.3. Agence de Régulation des Télécommunications 122
5.3.4. Cell phone operators 123
5.3.5. The judicial system and cell phone scams 124
5.3.6. Cell phone users and consumer associations 125
5.4. Socio-institutional governance of cell phone use in Cameroon: Optimal or approximate mediations? 126
5.4.1. Information deficit of the users. 126
5.4.2. Insufficient means of action. 127

5.4.3. Mis-selling of SIM cards by mobile operators: An “ingredient” of mobile scammers 128
5.4.4. The ease of monetary transactions 129
5.4.5. Technological constraints and border porosity 129
5.5. Conclusion 130
5.6. References 131

Part 3. Digital Risks: Practices and Mediation 135

Chapter 6. Towards a Normative Prescription of Information Practices on Digital Social Networks: A Study of Documentary Pedagogical Projects in Middle School 137

Adeline ENTRAYGUES

6.1. Introduction 137
6.2. Contextualization of risk 138
6.3. Issues to consider 138
6.4. Research objects 139
6.5. Research protocol 142
6.6. Risk regarding DSNs in the pedagogical approach 144
6.6.1. Raising awareness of risks: An obvious approach for teacher librarians 144
6.6.2. Considering the views of learners and teachers 145
6.6.3. Considering the risks: Learners aware of digital dangers 148
6.7. Discovering DSNs in a school context: Dealing with risks 151
6.7.1. Pedagogical projects on DSNs to prevent risks: Teachers’ perspectives 151
6.7.2. Overcoming risks: Learners’ perspectives 152
6.8. Perspectives for an information culture 153
6.8.1. Risks, standards and education 153
6.8.2. A culture of information in training 154
6.9. Conclusion 155
6.10. References 155

Chapter 7. MIL as a Tool for Teachers to Prevent Risk and Transmit Digital Culture 159

Julie PASCAU

7.1. Studying digital technology in schools from the perspective of teachers’ representations 159
7.1.1. Why be interested in representations? 161
7.1.2. The social representation of digital risks through the analysis of institutional discourses 163

7.2. What do digital and media literacy evoke in teachers? 164
7.2.1. The weak presence of digital technology and MIL in elementary school 165
7.2.2. Risks in the representations of MIL among primary school teachers 166
7.2.3. A positive perception of the role of digital technology in the classroom 169
7.3. The contours of media and information literacy according to teachers 171
7.3.1. The objects of MIL from the discourse of primary school teachers 172
7.3.2. What does digital technology mean for teachers? 173
7.4. What does the requirement to transmit digital culture mean for teachers? 178
7.4.1. Digital culture: A very vague concept. 178
7.4.2. What primary school teachers think digital literacy means 180
7.5. Conclusion 187
7.6. References 189

Conclusion. 193
Camille CAPELLE

Postface. 197
Vincent LIQUÈTE

List of Authors 201

Index 203

Foreword

A complex and indeterminate digital environment brings uncertainty to our everyday decisions and the risks that can result from them. Resolving this uncertainty, and adapting to an environment that bears different risks, questions the very perception we have of it. In the world of education, the way we learn to learn in an open digital world and the risks we encounter are orchestrated. Concerns about digital risks have come to constitute the cornerstone of our relationship to information and relevant educational issues. The risks that teachers and students associate with their information environment, the pitfalls encountered and the strategies proposed can be examined. Between aversion, avoidance, recommendations or resolution, a "relationship to risk" – a specific educational approach – is constructed in the informational and professional sphere. A "notion" of risk, which goes beyond the risks themselves, has become an approach associated with and accompanying digital practices in education. The need for understanding is associated with vigilance and the project of teaching informational risks. Research on digital risks in an educational and professional context is as much about the risk paradigm as it is about the informational uses associated with how they are perceived.

The notion of risk

The risk model brings to the fore the perception of an event that may occur, in "a future that may happen" (Beck 1986), which is not fully within our control. This event is sensed (and is not yet happening or real) as a danger that we must anticipate by picturing it and taking precautionary measures. This becomes a reverse causation of our action. Paradoxically, we tend to ask ourselves: how can we avoid the risks (even though it is the awareness of the risk which can ward off danger)? The perception of risk(s) then becomes a situation whereby the activity is associated with its uncertainties. Risk is distinct from danger, whose probability it relates to; it is distinct from the sole threat insofar as it is nourished by the perceptions that we have of it. Thus, for Ulrich Beck, at the heart of a civilization of risk, "risk has become the measure of our action"[1]. He describes the evolution of human and political action, linked to "scientific–technical–economic modernization", as a response to the disorders resulting from its development (industry, environment, economy, etc.). Risk does not designate danger; it constitutes the answer in line with an analysis or a preventive action. It becomes a model for the development of human conduct, a reason for action. For Beck, risk is integrated into a positive culture. With reference to "emancipatory catastrophe", he calls for new regulations of the human mind by making it the principle of an anticipatory consciousness and of an action that supposes taking risks. "Risk", for Beck, is not only the danger but also its perception and a critical modality that calls for a positive culture of change and a reflexive steering of action. Digital risks, associated with the socio-technical modernity described by Beck, are new challenges that require the construction of "new regulations for action": for Beck, this is a "disruptive innovation". Major cultural transformations are taking place, such as that of digital activity, along with its uncertainties and possibilities.

Digital risks and the context of education

What do we mean by "digital risks"? The expression brings into play both the informational character of human activities, its technological and social space, and the human factors driving the activity. They concern the informationalization of human activities and the relationship to the information regime. At different scales, micro (that of the digital subject), meso (that of the learning devices) and macro (that of the generalized register of interactions), it puts a strain on informational ecology. Digital risks now belong to both the generalized digital transposition and the activity itself that unfolds in the uses, their "sociodynamics" (according to A. Moles' expression), where there are close links between social issues and knowledge.

1 Beck, U. (1986). *La société du risque : sur la voie d'une autre modernité*. Flammarion, Paris.

The recognized and stated risks overlap with a multidimensional set of different registers encountered in an informational ecology: "the technical risk (computer security); the social (equality, safety) and political risk (security, indoctrination); the cognitive and psychological risk (educability); the ethical and legal risk (respect for rights, digital identity, e-reputation, but also online harassment related to this practice); the risk related to health (static activity, caution with regard to radio waves, lack of sleep due to blue light), etc."[2]. The informational risk is more often associated with informational effects in the sphere of knowledge, economy, political information, data use or an ecology of attention[3]. The plurality of the stated risks is to be compared with the global modality of risk in our modes of thinking and acting in a digital context. Each of the risks identified is part of a context of use; it refers to a specific link between perception and activity. For example, the informational scenarios stage the construction of knowledge and integrate the perception of the different risks mentioned by the contributions to the work into the uses. But, through these contexts, risk posture, its perception, the conduct of the activity, from prevention to risk-taking, make it, beyond the simple attribution to the digital of the risks incurred, a reflexive and responsible occurrence of each person's activity.

The educational context, between mediations and practices, creates a particular tension. In the relationship between teaching and learning, education encounters the risks brought about both by technological and social transformation and by human activities that have moved or developed in the digital world. How can we escape a paradoxical and antagonistic culture, between the need for digital technology to teach and learn on the one hand, and on the other, an overvaluation of the risks that would problematize its use? How can we build a "learner" autonomy? How can we compensate for the ethical fragility of digital technology, which seems to escape older paths of knowledge? How do teachers (and students, in their "profession") reconcile their personal and professional practices (Capelle op. cit.)? How can the desire to teach turn into the risk of teaching (Cordier 2017)?[4] These questions, and others, lead us to a dialogue between practice and research in this book.

---

2 Capelle, C. (2018). Rapport final de projet de recherche eR!SK – risques numériques et école 2.0. Research report, IMS Laboratory, University of Bordeaux, Bordeaux.

3 Citton, Y. (2017). *The Ecology of Attention*. Polity Press, Cambridge.

4 Cordier, A. (2017). Les enseignants, pris dans des injonctions paradoxales. Hermès, La Revue, 78(2), 179-186.

Risk perception

How do we represent risks in our minds? The "existence" of risk can be split up into the real danger perceived and the representation we have of it. Between real and perceived risks, real and imaginary, for all the actors – in this case, teachers and students – the risk incurred (factual aspect) is as important as the perception of the digital "cause" linked to it. The postures engaged find their reason: avoidance, risk literacy, the staging of activities that ensure awareness (fact-checking), reflexive practices, etc. All of the actions undertaken, the literacy of the risk and the awareness of the "cause" of the risk, have to be taken into account. All of the actions undertaken, and the articulation of risk, are based on a perception that contributes to its reality. Representations thus play a strategic role in prevention and in the capacity of individuals to conceive and define risk. The meaning given, what "we think we are doing" (in the sense of Bruner[5]), brings together the inner and outer facets of the situation. The perception of risk does not necessarily conform to its reality; but it is the reality of the action it supports. The representations (individual and shared subjectivation) interfere with the uses as much as those which construct them. The represented mode belongs to the work; it becomes a critical point. Articulated with reality, integrated with a reflexive thought, the perception of the risk becomes a lever for our practices.

Those who act in a digital environment operate on different beliefs and hypotheses. They concern the conditions of use and the perception of possible perils. The contributions in the book focus on the perceptions and postures involved in professional and/or learning activities, on the perceptions of the actors involved in learning and teaching activities. It is described by teenagers as "a big scary nebula" (Cordier op. cit.). For some, the risks can be considered as negative or potentially dangerous effects of informational mediations: for others, the digital environment offers a resilience to risk. The attribution of the amplification or reasoned treatment of risks to digital technology calls the meaning of education into question. But are we speaking the same language? A report published in 2008 by Christine Dioni[6] entitled "The student's job and the teacher's job in the

5 Bruner speaks of implicit theories: "It is not enough to describe what the child is doing: we must be able to determine what they think they are doing and why". Bruner, J. (1998). *Le développement de l'enfant : savoir-faire, savoir dire*. Presses Universitaires de France, Paris.

6 Dioni, C. (2008). *Métier d'élève, métier d'enseignant à l'ère numérique : rapport de recherche pour l'INRP*. INRP, Paris.

digital age" emphasized the issues of discrepancy between the perceptions of students and those of teachers, as well as between the perceptions and reality of practices. The student's mirror is the "school's" perceptions of digital technology, which do not necessarily correspond to those of their peers. The expression of teenagers' experience becomes a driving force for the necessary reflexivity for practices. The teachers themselves keep personal practices and professional practices at a distance[7]. Perceptions of the digital world and its uses are thus linked to the informational experience.

Educational response and research

Risk has become a social and informational reality, both objective and subjective, "that drives us to act". Its integration as a regulated condition of the informational experience supposes an intelligence and an experience of the activity shared with the students. The integration of the risk dimension in teaching and learning activities is associated with "media literacy" in our educational system. Its objective is the awareness and management of students' activity in an information environment. The ability to assess digital risks and the possibility of managing events/information in a critical way are the horizon. Media education opens up, beyond the mastery of informational tools, by taking into account the fragility of information systems or illicit uses, to the informational experience and to an information culture. It is not a matter of protecting ourselves from risk, but of designing support for students that is adapted to information contexts. "Risk education", integrated into teaching and learning, examines the new skills required for an information society, as well as pedagogical mediation. Risk education integrated with information literacy responds to the need for the new regulations (Beck) required for an information society. Different paths, the plurality of situations and uses, are thus to be explored in the educational field. The notion of risk, as a condition, is not a limit to be fixed but a critical tool within learning.

This book brings together various contributions at the intersection of professional action and research. The proposed analyses are in line with the second edition of the conference "Knowledge and Information and Information in Action (CIA)" entitled "From risk perceptions to action in a

7 Capelle, C. and Rouissi, S. (2018). Représentations et stratégies de jeunes enseignants face aux réseaux sociaux. *Les Cahiers du numérique*, 14(3–4), 13–34.

digital context" held on April 3 and 4, 2019 at the University of Bordeaux, France, organized around the project eRISK (Education 2.0 and "digital" risks; representations and pedagogical practices) led by the RUDII-IMS team[8]. It proposes a renewed focus on the perceptions of risks in the educational and professional context. Different paths for research and action are presented: those of the subjective dimension of risk perception in the shaping of practices and the construction of action strategies in a situation[9]; or those of objectivizing analyses of an information ecology (arena) and the organization of action devices. The epistemological limits of information sciences are questioned as much as the uses and practices.

Risk belongs to the cultural transpositions linked to digital technology. It becomes a principle of activity in an uncertain world, a critical and social component in support of education. This research contributes to its understanding and its reflexive inscription in information, "academic" and societal behaviors.

Franc MORANDI
September 2021

---

8 RUDII: Représentation Usages et Développement des Ingénieries de l'Information, IMS: laboratoire de l'Intégration du Matériau au Système (IMS, CNRS UMR 5218). CNRS, l'Université de Bordeaux, INSPE d'Aquitaine (2016–2019). *Projet eRISK: "Risques numériques et école 2.0 Éducation des jeunes générations et responsabilité des enseignants"*. Hypothèses [Online]. Available at: https://erisk.hypotheses.org/category/projet.

9 Lave, J. and Wenger, E. (1988). *Situated Learning. Legitimate Peripheral Participation*. Cambridge University Press, New York.

Introduction

In recent years, the notion of digital risk has invaded the media. Numerous publications suggest that digital technology is a source of all sorts of threats, whether at the informational, psycho-social, ethical, cognitive, health, technical, socio-economic, legal or environmental level. For example, we can cite the denunciation of the risks of addiction and "techno-addiction" linked to screens among young people[1] or the impoverishment of reading practices[2] and thought[3]. This risk-based approach tends to solidify the description of individual and social practices in categories that do not reflect their evolution and diversity, nor the capacity of actors to innovate, which implies taking risks.

The intention of this book is to go beyond a stigmatizing approach to risk. Instead, it is to consider it as the center of discourses, representations and practices. It is necessary to deconstruct and understand it in order to propose tailored support.

According to Ulrich Beck, "risks designate a future that must be prevented from happening". Real and imaginary at the same time, they are, in his view, "an event that has not yet occurred which motivates action"[4].

Introduction written by Camille CAPELLE.

1 Lardellier, P. and Moatti, D. (2014). *Les ados pris dans la Toile : des cyberaddictions aux techno-dépendances*. Le Manuscrit, Paris.

2 Carr, N.G. (2011). *Internet rend-il bête ?* Robert Laffont, Paris.

3 Desmurget, M. (2019). *La fabrique du crétin digital : les dangers des écrans pour nos enfants*. Le Seuil, Paris.

4 Beck, U. (2008). *La Société du risque : sur la voie d'une modernité*. Flammarion, Paris.

This "growing appearance of risk in the world" is also denounced by Patrick Peretti-Watel. For him, risk "is a danger without cause, a damage without fault, which nevertheless becomes predictable and calculable"[5]. The perception of digital risks therefore motivates individuals to act, but how? And in what ways? How do they critically assess their practices?

The term digital risk is often used to designate the dangers linked to the security of computer systems, particularly in computer science and management science. The field of law is also concerned notably around the problems of information security, personal data protection[6] and e-reputation[7]. Other publications also concern ethical issues, in sensitive areas of personal information protection, such as health.

In psychology, digital risks are treated from the point of view of cognitive overload[8] and addictions[9]. The notion of informational risk is often linked to knowledge management[10], problems of manipulation and critical evaluation of sources[11] and information asymmetry[12]. For the manager of a company, the digital risk is linked to the security of data, strategic information or the reputation of the company.

With regard to education and the perception of risks associated with the use of digital technologies, a report published by Christine Dioni in 2008, (The student's job and the teacher's job in the digital age), emphasized the problems of discrepancy between the perceptions of students and teachers, as

5 Peretti-Watel, P. (2010). *La société du risque*. La Découverte, Paris.

6 Rouvroy, A. (2014). Des données sans personne : le fétichisme de la donnée à caractère personnel à l'épreuve de l'idéologie des Big Data. *Étude annuelle du Conseil d'État : le numérique et les droits et libertés fondamentaux*, La Documentation française, Paris.

7 Du Manoir de Juaye, T. (2014). Le risque informationnel au filtre du droit. *Documentaliste – Sciences de l'Information*, 51(3), 37–40.

8 Tricot, A. (1998). Charge cognitive et apprentissage : une présentation des travaux de John Sweller. *Revue de Psychologie de l'Éducation*, 3, 37–64.

9 Blaya, C. (2015). Les jeunes et les prises de risque sur Internet. *Neuropsychiatrie de l'enfance et de l'adolescence*, 63(8), 518–523.

10 Robert, P. and Pinède, N. (2012). Le document numérique : un nouvel équipement politique de la mémoire sociale ? *Communication et organisation*, 42, 191–202.

11 Serres, A. (2012). *Dans le labyrinthe : évaluer l'information sur internet*. C&F, Caen.

12 Pariser, E. (2011). *The Filter Bubble: How the New Personalized Web is Changing What We Read and How We Think*. Penguin, London.

well as between mutual perceptions and the reality of practices[13]. Just over 10 years later, what is the situation now?

The stakes of digital practices in terms of learning[14] and the socio-economic risks linked to the digital divide[15] lead the actors of education to ponder effective means of confronting them. The idea of media risks is also used to denounce inappropriate content, such as violence, incitement to hatred or pornography on the Internet, calling for the establishment of standards[16]. The risk of ideological and political manipulation is currently back in the media spotlight through the analysis of conspiracy theories, the topic of "fake news" and young people's susceptibility to propaganda.

For the educator, risk is also linked to the illicit use of information by students, in particular, the use of information that falls under law (law on the freedom of the press of 1881, law of July 13, 1990, known as the Gayssot law, prohibiting the dissemination of information of a revisionist, racist or sexist nature, or that promotes intolerance, for example). Many other uses can be problematic, such as disrespecting others, the privacy of teachers, education staff or other students. So-called "cyberbullying" situations via Web 2.0 tools are also problems that educators have to deal with (messages or video broadcasting via personalized and live channels on Snapchat or Periscope, for example). The digital practices of young people are all the more complex as they mix creativity and skills development, intimacy and self-exposure. What these multidisciplinary works have in common is that they seek to identify what digital technology does to the individual and to society in different contexts (educational, professional, political, etc.). The objective of this book is to compare the discourses on risk with the practices and representations of the actors in different professional fields and in the educational field, and subsequently, to explain their stakes in a digital context.

13 Dioni, C. (op. cit).

14 Jehel, S. and Saemmer, A. (2017). Pour une approche de l'éducation critique aux médias par le décryptage des logiques politiques, économiques, idéologiques et éditoriales du numérique. *tic&société*, 11(1), 47–83.

15 Plantard, P. (2011). *Pour en finir avec la fracture numérique*. FYP Editions, Limoges.

16 Jehel, S. (2011). Contenus médiatiques à risque et construction identitaire des préadolescents. *Sociétés et jeunesses en difficulté. Revue pluridisciplinaire de recherche*, 11.

Our assumption is that individuals are armed with resilience[17] in a changing world. The perception of digital risks can be a resource or a lever to develop a critical culture in a digital context[18]. The contributions in this book present the means we have today to move from risk perception to risk information, from information to knowledge, and from knowledge to action[19]. They will be of particular interest to teachers, trainers, mediators, information and media professionals.

The book is divided into three sections. The first section deals with the perception of risks in education and shows how the Internet currently constitutes an ecosystem in which the relationship between the teacher and their students is inevitably positioned.

Chapter 1 shows the impact of alarmist discourses on educational practices through a study on the representations of digital risks among young teachers. These discourses seem to have an important resonance with teachers when they start their career and can dissuade them from implementing any digital practice at school with students.

Chapter 2 focuses on adolescents' perceptions of fake news. The study highlights variations in reflexivity towards critical evaluation of sources and disinformation across different age groups of students.

Chapter 3 aims to better understand the pedagogical positions and choices of teachers, particularly in the area of Media and Information Literacy. Based on a survey conducted among teachers, the representations and non-formal practices identified lead the author to advocate for a cultural approach to digital education in professional training.

The second section deals with informational risks, as well as economic and social risks in professional environments. It focuses on the practices used to deal with these situations.

---

17 Tisseron, S. (2013). Résiliences : ambiguïtés et espoirs. *Annales des Mines-Responsabilité et environnement.* ESKA, 17–21.

18 Capelle, C., Cordier, A., Lehmans, A. (2018). Usages numériques en éducation : l'influence de la perception des risques par les enseignants. *Revue française des sciences de l'information et de la communication,* 15.

19 Liquète, V. (2011). *Des pratiques d'information à la construction de connaissances en contexte : de l'analyse à la modélisation SEPICRI.* University of Rouen, Rouen.

In Chapter 4, the authors look at information risk from the point of view of business leaders. They show that this notion is broader than digital or computer risk, which is restricted to digital media. The risks mentioned by managers in the context of their security policy include many informational risks linked to other media.

Chapter 5 highlights the difficulties faced by organizations in Cameroon in dealing with the risks of scams and fraud in the use of cell phones by the population. Despite attempts at prevention by institutional organizations and cell phone operators, users suffer from an information deficit and the lack of means of action is not conducive to optimal mediation of these risks.

The third section looks back at the practices and mediations around digital risks in education. In particular, it examines the contribution of Media and Information Literacy, as implemented in schools in France.

Chapter 6 shows that the notion of risk is often the starting point for pedagogical practices concerning digital uses and appears to be deeply rooted in the teachers' vision of information culture. Although school learning practices are still quite far removed from the private sphere, we can see that teachers in the field are concerned with developing knowledge in students that they can transfer from one context of use to another.

Finally, Chapter 7 compares the discourse of teachers in primary education with institutional texts concerning Media and Information Literacy. This comparison shows that the development of digital culture advocated in institutional texts remains a vague subject for primary school teachers. Beyond the operational approach to the use of digital tools, it is confirmed that primary school teachers lack training in the cultural approach to the web and its media.

PART 1

Risk Perceptions, Education and Learning

1

Digital Risks: An Obstacle or a Lever for Education?

1.1. Introduction

Given the challenges of new technologies that society must face, the notion of risk has become a difficult one to grasp, because it refers to dreaded events and individuals' worries and fears. For the sociologist Ulrich Beck (2008), risks no longer come only from outside (natural risks). Modern society, with its technological progress, in particular, generates risks in the sense that it debates them abundantly and seeks to prevent them in increasingly optimal ways: this is the emergence of a "risk society". The introduction of digital technology into schools and families has generated numerous myths (Musso 2008; Amadieu and Tricot 2014) and unfounded fears (Cordier 2015; Plantard 2016) that are also widely debated. Digital refers in discourse both to computer tools or techniques (Internet, computers, smartphones, web platforms, etc.) and to uses. Indeed, the digitization of many practices has increased the number of situations in which digital tools are used, be it for information, communication, transmission and so on. Students, as well as teachers starting their careers, are now part of a generation that has grown up with the evolution of digital technologies and has built its practices and representations around them.

This leads us to examine the conceptions of digital technology among this new generation of teachers and the status they grant to digital literacy. In

---

Chapter written by Camille CAPELLE.

For a color version of all the figures in this chapter, see: www.iste.co.uk/capelle/digitalrisks.zip.

this chapter, we propose characterizing different types of digital risks and subsequently compare them with the perceptions of teachers at the beginning of their career. The objective is to understand how these representations around the notion of "digital risks" can be a brake or a lever for educating young people about digital issues.

This text builds on the research of the eRISK project, Digital Risks and Education 2.0, which was conducted between 2016 and 2019 with the support of the Maif Foundation.

First, we will define the notion of "digital risk" and characterize the diversity of risks that may correspond to this notion. We will then see which risks school faces with digital technology. We will then present the methodology used to collect the representations and declared practices of new teachers regarding these risks. Finally, a third section will be devoted to the analysis of the teachers' representations and practices and to a reflection on the impact these have on the education of students in the digital world.

1.2. Digital risks and education: what are we talking about?

1.2.1. *Digital risks*

Digital risks can be considered threats, of varying levels of danger, which can manifest themselves during or following a digital activity and which are likely to affect the user or to have harmful consequences for others. These risks cover a wide variety of domains that we have grouped into nine types, following a preliminary study carried out collectively within the framework of the eRISK project by means of a thematic exploration in the major media. This step allowed us to develop this typology, which we then took a step back from and backed up with the scientific literature. We have thus identified technical, informational and political, cognitive, psychosocial, health, socioeconomic, ethical, legal and ecological risks. Here, we propose to briefly characterize these different types of risks.

Technical risks

Technical risks are related to potential malicious intrusions likely to damage digital equipment in one way or another (computer, tablet, smartphone, USB key, CD-ROM, etc.) and spread to other equipment. The motives behind these attacks can vary: to destroy documents to harm an individual or a group of individuals (economic, military or governmental organization, etc.); to infiltrate an information system to steal information with the idea of denouncing or conducting industrial espionage, for example; or simply with the idea of taking up a challenge. We use the term hackers to designate those individuals who illegally act on other computer systems. These acts fall under cybercrime and are punishable by law.

Informational risks

Informational risks can be very diverse in nature. They can involve the difficulty of knowing the reliability of certain information on the web, due to the proliferation of falsified information (rumors, disinformation) disseminated voluntarily and with the aim of causing harm. The risk of being informed on the Internet is also that of being conditioned by the giants of the Web (GAFAM), who direct individuals according to their profile towards information intended to "correspond" to them. This phenomenon, exposed by the idea of filter bubbles (Pariser 2011), stems from the fact that on the search engines or platforms used, the algorithms that filter and select information can be based on the users' profile (from their personal data, their search and browsing habits); their network of relationships (i.e. relationships that generate frequent interactions and are considered more influential by users); the popularity of information (popularity based on the number of clicks, retweets on social networks, likes, etc.). The manipulation of information on the Web also manifests itself through "conspiracy theories" (Bronner 2013) that are disseminated in order to generate doubts about certain scientific facts or political powers. In this way, ideologies can be relayed and generate socio-cultural or political conflicts.

Cognitive risks

Cognitive risks refer to the disruption of attentional capacities that can be generated by intensive use of digital tools or cognitive disorders. In 2011, the American writer Nicholas Carr denounced the Internet in a book entitled "The Shallows" on the grounds that it would lead to an impoverishment of reading practices and thinking. Researchers in cognitive psychology (Amadieu and Tricot 2014) have shown that reading skills on the Web are indeed different from those required on paper, a thesis confirmed by the American neurolinguist Katherine Hayles (2016), who explains that the human brain has adapted to digital environments to acquire a new mode of attention: hyperattention. This new form of attention allows us to find our way around web content, through hyperlinks.

For some, this evolution in the ways of being attentive, with varying levels of concentration, of retaining certain information to varying degrees in various daily activities would be the harmful consequence of an overexposure to screens. For others (Citton 2014), the cause would rather be sought in the marketing model of platforms that seek to capture monetizable attention above all.

Psychosocial risks

Among the psychosocial risks, exposure to shocking, hateful or pornographic content (Jehel 2015) on certain websites or in games can generate psychological and social difficulties, particularly during adolescence. The phenomenon of screen addiction is also denounced on the basis that it distances some users from everyday social activities, to the point of isolation. Even if the origin of the malaise of these individuals comes, most of the time, from the personal and family context, the digital environment can reveal these difficulties (Stora 2018). Verbal violence or humiliation repeatedly suffered through the Internet (Blaya 2013) are also threats likely to affect the psychological and moral health of individuals. We use the term cyber harassment to define these practices in which aggressors act with a feeling of total impunity, believing themselves to be hidden by the anonymity of the Internet.

Health risks

Among the health risks worth mentioning are the potentially harmful effects of screens on eyesight, which can lead to visual fatigue as well as sleep disturbances. Posture disorders linked to digital uses or exposure to radio waves also generate concerns, as conveyed in speeches and in the media.

Socio-economic risks

Socio-economic risks correspond to what has been called the fractures linked to inequalities among users (Plantard and Le Mentec 2013). These fractures concern different levels: access to the Internet or to digital equipment (equipment inequalities); the urban or rural living environment (access inequalities); the environment and the socio-professional category of a user's parents or of an individual (usage inequalities); the support available to users and their ability to train themselves (usage performance inequalities) (Ben Youssef 2004)

Ethical risks

Ethical risks can refer to the use of personal data and tracking during web browsing. Even if such use is not always harmful to the user, they are rarely well informed about the nature of the personal data collected or the ways in which it might be used. In addition, the security of this data is not always fully ensured and poses problems regarding the protection of privacy on the Web and respect for others (Merzeau 2013; Rouvroy 2014, pp. 407–422; Cardon 2015).

Legal risks

The legal risks on the Internet are complex given that the application of the law has been adapted to technological evolutions and the use of the Web. If the law applies on the Web as in everyday life, digital spaces generate new cases that require necessary adaptations. In particular, the risks concern the violation of intellectual property, personal data, image rights or even identity theft. These infringements are punishable by law and can be subject to fines and prison sentences.

Environmental risks

The environmental risks linked to digital uses are less related to the current consumption of energy for operating, storing and recharging electronic devices than they are to the pollution generated during the manufacturing of the equipment: the lithium in the batteries, the tantalum in the electronic cards, the indium in the screens and other components are produced far from the assembly sites and require a lot of energy to assemble them. These are also scarce resources that are being depleted at an increasing rate. The fact that this equipment is not recyclable also has a major ecological impact. The speed of technological advances and the obsolescence of digital equipment lead to the production of more and more waste which is not treated in a responsible way.

Box 1.1. *Typology of digital risks*

This diversity of digital risks is reflected in the media and societal and scientific discourses, thus fueling the debates. Often presented in an alarmist way in the media, these speeches alert users to dangers that are likely to have real impacts on their daily lives, often playing on the register of fear. However, while the stakes around digital technology are real and deserve to be considered, the probability of being personally affected by one of these threats remains low and occurs only in very specific cases. These discussions on risks are, however, likely to have a particular resonance with teachers, parents, educators and, whether directly or indirectly, with teenagers.

1.2.2. *What are the risks in education?*

For several years now, the educational system has been communicating on the need to prevent certain abuses such as cyberbullying, violation of image rights or disinformation. Many actors work closely with schools to produce and disseminate information to members of the educational community. The *Commission Nationale de l'Informatique et des Libertés* (CNIL) works on the rights and duties on the Web and the protection of personal data; the *Centre de Liaison pour l'Éducation aux Médias et à l'Information* (CLEMI) assists in the analysis of sources and the critical evaluation of sources; the police or associated organizations are also involved in informing and supporting the actors of the educational community, who are faced with digital risks.

Those critical of digital technology in schools argue that it is likely to increase inequalities among students. They consider that the digital practices of young people are essentially consumer or leisure practices, far removed from academic expectations. Thus, working at school with these tools would likely present risks for students such as cognitive overload and attentional deficit, and could reinforce certain stereotypes related to the heterogeneous support of families on the Internet, which does not allow all students to have cultural references or sufficient critical distance from the content available. According to this presupposition, which is still widely criticized, "a large proportion of young people would thus fall prey to commercial or ideological manipulations of all kinds and could in no way take advantage of the potential for emancipation and access to culture offered by digital technology…" (Becchetti-Bizot 2017).

On the other hand, the school system has been trying for years to set up several plans to develop digital uses, including the computer plan for all in 1985, the digital plan for education in 2015, and the digital plan for school confidence in 2018. However, digital practices are still not permanently embedded in school practices and the digital literacy of students is often a one-time practice that is not yet fully integrated in disciplinary teaching.

For Vincent Liquète and Benoît Le Blanc (2017), a "shift from comprehension to use or even manipulation" occurred between the 1970s and the 2000s with the arrival of digital technology in schools. Whereas audiovisual and radio technologies were established in schools from a comprehension and critical perspective, the integration of digital technology seems to be based on the need for technical mastery of these tools. This new approach could therefore run the risk of reducing students' ability to understand, analyze and criticize these information and communication technologies and their uses in society. The authors note a diversity in the ways in which teachers have appropriated and invested in digital technology over the course of this decade:

> [There is] the "prophet" teacher singing the praises of technologies and, more recently, the digital environment, the "sales" teachers wishing primarily to link the school to market trends and employability issues, the "activist" teachers proposing alternatives especially via freedom and the common good, and the "innovator" teachers claiming that digital technologies drive creativity and new ways of doing things (Liquète and Le Blanc 2017, p. 11).

Our hypothesis is that another mode of investment in digital technology by teachers has emerged in recent years in the face of perceived dangers for students or for the teacher themselves: that of renouncing any desire to integrate digital technology into the classroom, on the grounds of the risks incurred with respect to families, a feeling of incompetence or a certain moral code. This hypothesis led us to investigate in order to better understand teachers' feelings about these topical issues.

1.3. Questioning perceptions of digital risks among new teachers

1.3.1. *Why was this target audience chosen?*

Most of the new generation currently entering the teaching profession are "digital natives" (Prensky 2001), native to the Internet and digital tools. This generation uses digital technology on a daily basis to connect, communicate, inform themselves, entertain themselves, etc. These practices developed in the personal sphere are then likely to have an influence on their perceptions of digital technology and on their teaching practices with students.

We have chosen to focus on one category of teachers: that of new teachers with less than three years of experience in the profession. This choice seems judicious to us for several reasons. First, it is plausible that this new generation, which has grown up with digital technology, could be the bearer of a new, more realistic view of digital technology, against the idea of a certain utopia that the Web initially embodied for many. Secondly, teachers who are starting out in the profession are committed to meeting institutional expectations and are therefore expected to carefully consider the official texts that promote this education in and through the digital world. Thirdly, teachers who are just starting out have high expectations in terms of training and are likely to bring to light shortcomings and needs, thus leading to a re-evaluation of teacher training.

1.3.2. *Methodology and data collection*

The survey conducted as part of the eRISK project took place between 2016 and 2019. Based on the typology of digital risks presented above, the survey explored the reported representations and practices of new teachers. The survey questionnaire, entitled "*Enquête sur les pratiques numériques des enseignants*" (survey on the digital practices of teachers), was distributed to all elementary school teachers and secondary school teachers, as well as principal education advisors in the academies of Bordeaux and Créteil, regardless of their seniority level. A total of 3132 teachers felt that this survey concerned them, and they responded in full. Among these 3132 respondents, 724 teachers were new teachers (i.e. the target of the study). We therefore analyzed the responses of these new teachers exclusively, with the idea of also being able to compare these results with those of more experienced teachers. This work is not covered here and may form the subject of a subsequent analysis.

It should be noted that the criterion of the number of years of experience is independent of the age of the teachers. Indeed, our panel of teachers interviewed was mostly under 35 years of age (72.4% of them), but the rest of the teachers were older, with most of them having changed careers. Following an initial analysis of the results, semi-structured interviews were conducted with 15 volunteers from the teachers who responded to the survey. The objective here was to present the major trends of the study to them in order to get them to react to them and explain their positions. This mixed data collection, both quantitative and qualitative, made it possible to refine and better understand some of the survey's results, considering the diversity of representations around this subject.

The perceptions of digital risks were examined through the positions and discussions of the teachers interviewed. They relate both to their takes on their personal, often daily digital practices, and to their takes on their role as educators of young people. In order to see what conceptions of digital technology exist among this new generation of teachers and what status they afford to digital literacy, we sought to understand the diversity of these practices while aiming to identify sets of positions governed by common values, beliefs or principles.

1.4. Teachers' perceptions of digital risks

Most of the teachers interviewed in the survey claim to use digital technology in an educational context: to prepare their lessons (93.9%), to communicate by e-mail with their colleagues (90.3%) and to illustrate their lessons (85.5%). The use of digital technology is therefore mainly for professional purposes, but is far less concerned with educating students in digital technology; only 49.4% declare that they train their students in digital technology.

By analyzing the interviews, three main sets have emerged that characterize the teachers' discussions and positions:

– the first corresponds to teachers who renounce any practice of digital technology in the classroom, in particular because of certain perceptions of risk;

– the second concerns those who restrict the field of use for fear of being confronted with certain risks;

– the third group includes those who consider digital risks in the exercise of their profession and who seek to prevent them through education.

1.4.1. *When perceptions of risk inhibit any practice*

Certain perceptions of digital risks seem to be an obstacle. The use of digital practices with students is perceived by some as a problem factor, especially in their relationships with their parents. A schoolteacher explains his reluctance to have his students work on the Internet to us:

> I'm not convinced of the parents' response and I'm afraid of being annoyed, frankly, by the parents' response. So, I've never used it and so, yes, that's it (schoolteacher, 41 years old).

In secondary education, the use of social media platforms (including YouTube, social networks or blogs) is largely avoided, even though they constitute an important angle of media and information literacy. One schoolteacher expresses her difficulties in considering solutions for covering media and information literacy with her students:

> It's true that it's very, very regulated, so, uh, even the blog, we're not going to use just any blog, so it's the academic blog and we're obliged, it's very formalized. So, it's true that I know that it can be a hindrance, yes (fourth-grade schoolteacher, 25 years old).

This avoidance is dependent on their risk perception and on a kind of ethical principle that they apply to themselves as well as to the students:

> I can't get into it, first of all because I was probably reluctant to it, and the dangers it can bring. I saw when I was working in secondary school, we saw students' Facebooks and what they put on them, and it's unbelievable... So, for me, it's more of a danger than a tool, in fact (teacher in pre-kindergarten, 44 years old).

An English teacher who is committed to educating students digitally explains:

> There are many teachers for whom using digital networks, even Twitter, is too scary (English teacher in a high school, 25 years old).

Social networks are indeed perceived as spaces that must remain totally inaccessible at school, associated with "dangerous sites" and treated in schools, according to teachers, on the same level as "pornographic sites":

> When they go on the Internet, there are a lot of filters that prevent them from accessing a certain number of so-called dangerous sites. I don't know how to talk about it... well, I don't know how to talk to the students about these... risks, I mean. In fact, when they go on the Internet at school, they don't

> have access to social networks, pornographic sites, sites… I don't know, maybe dangerous sites, or so-called dangerous sites, in the educational world (art teacher in a high school, 29 years old).

We also see that the risks evoked for students are strongly influenced by their personal experience and the concerns they may have for themselves, such as here, on social networks:

> But the problem with social networks is that they are very… it's a bit of a vent for everyone and I don't really like that system (management teacher in a senior high school, 39 years old).

These concerns crystallize around the use of social networks and the information disseminated or used on the Web. They are first projected onto the students, as expressed by this English teacher:

> I just think that there is no limit in fact with the digital environment among students… and it goes over our heads. It goes over our heads, their heads and the parents' heads, over everyone's and uh… at the same time it's addictive for them so… it's addictive and it's destructive at the same time and it's risky, too. That's why it's really worrying (English teacher in a high school, age 25).

The cell phone is also a subject that gives rise to debate and apprehension within schools. Its prohibition in schools and colleges, apart from educational uses, has undoubtedly reinforced certain apprehensions among teachers, as expressed by this high school teacher:

> Digital technology, for me, is something… I don't know how far it can go. That is to say that… if I do a class with students, for example, and I ask them, which won't happen, to use their cell phones, so there are 24 students or 28 students, since these are classes of 24 or 28, if I find myself with two, three students using their phones for purposes other than pedagogical purposes… not only will I not have the means to block them, I'll have no way of knowing where it went, how it was broadcast or anything. Or what was broadcast. I won't know

> until later, or by word of mouth. So that's it, really, the fear of that. The fear of being filmed without one's knowledge, the fear of that being sent on the Internet, on any type of social network it can be posted on and not having control over this post (aeronautics teacher in a high school, age 55).

The feeling of losing control over certain misuses by students is clearly an obstacle to digital literacy. Teachers feel that digital practices put them at risk and cannot work out how to avoid these misuses or educate students on responsible use.

Through testimonies, we can experience the complexity of this subject, the impact of depictions of digital risks and the role they play on emotions: fear of wrongdoing, fear of not knowing how to teach digital literacy, fear of generating problems for students or of encountering problems with students themselves.

1.4.2. *When perceptions of risk freeze practices*

For a large majority of teachers, digital technology is not just dangerous. Many of them perceive the pedagogical advantages of certain practices with students, but do not necessarily follow through for various perceived reasons: the procedural heaviness of complying with institutional expectations, a lack of resources, meaning that not all students can use the technology and a lack of time to respond to all the ministerial injunctions in the first years of practice (program, education, training, etc.).

Generally speaking, digital technology in schools is also synonymous with promises that are proclaimed and not always kept:

> What intrigued me is that the theory is great but the practice is quite complicated, because in fact we don't have the means on site to implement what we would like to be able to do (fourth-grade schoolteacher, 23 years old).

And those who try the experience do not necessarily do it again:

> We had worked on the method and so on, and then the computer system crashed, as they say, so we lost all our work, but for me it is not only the technical aspect that slows

> everything down, but actually the procedural and almost institutional aspect too. We are asked to operate in a very precise way that does not necessarily correspond to the reality of the field and the needs of the students. Nor to the needs of the teachers either, for that matter (English teacher in a high school, age 25).

However, beyond operational constraints and feasibility for implementation, risk perceptions also influence the frequency that digital media is used in the classroom, as expressed by this mathematics teacher, who feels that it generates social inequalities:

> Everything that concerns the use by the students is excessively time-consuming, it requires an enormous amount of preparation time for an unsatisfactory result. Or else, for it to be satisfactory, it would have to be the only thing done, and this is absolutely impossible in terms of resources, in terms of time and in terms of interest… there are many disparities. Because the students don't necessarily have access to a computer at home, they don't necessarily have access to the Internet… and the work is not necessarily important enough for them to do it on their own (math teacher in a senior high school, age 31).

For this fourth-grade teacher, it is more the cognitive and psychosocial risks that generate apprehension:

> If it's to do a little bit of math practice etc., you know, at very specific times, then yes, but not so that it takes over because they have so many screens around them that over-stimulate them in the morning before going to school and then in the evening. I have the results of the number of hours spent on the screens, it's enormous and we see that they are tired, that they are agitated [...] society wants to show them video games, names of famous people, footballers, etc. When they write a story, they are not able to see the names of the people they are watching. When they write a story, they have trouble detaching themselves from that, and I'm a little worried about that, and I've been able to discuss it with older teachers, who agree without saying "let's go back to the good old days" and all that,

> but they've seen the evolution, and I spotted it after a year or so, right away (fourth-grade schoolteacher, 23 years old).

We note that these teachers very often evoke the practice and the manipulation of the tools by completely omitting the educational and cultural approach related to the digital environment:

> This approach to the use of digital tools is not yet something that is extremely valued from the point of view of entrance examinations to engineering schools (physics and chemistry teacher in a senior high school, 30 years old).

The training of students with digital technology even seems to be part of a prestige scale for teachers, in this case at the bottom of it, compared to what can be taught in the school curriculum.

Some teachers even consider that specific courses on digital uses should be given, and therefore almost excluded from the scope of their duties, in their view:

> I really deplore the fact that computer science is not a separate subject in France. Separated from technology, I mean. Or that there are many more hours of technology and therefore many more teachers and that there is a part exclusively on computer science (English teacher in a senior high school, 37 years old).

Another expressed the idea that students are not at the level to be able to learn with digital technology in their discipline, since they have not mastered the basics:

> I think we should even start by training them in the use itself, because ¾ of the students who are in high school, they don't know how to use Word or Excel properly, or the Internet either. All they know how to do is search in the space bar and go watch videos and stuff. They don't know how these things really work and when you try to show them, they say, "No, we know how to do it". But that's not true (English teacher in a senior high school, age 37).

There is a conflict among teachers between those, on the one hand, who perceive digital literacy as an accompaniment to a technical mastery of digital tools, and those, on the other hand, who envisage a cross-disciplinary and cultural education that covers all areas. We can feel the distress of these teachers who feel that they are at a dead end when it comes to integrating digital literacy into their teaching:

> Everyone does a little bit of what they want to do, or what they can do (math teacher in a senior high school, 49 years old).

Finally, it should be noted that half of survey respondents (52.8%) stated that they do not carry out digital risk prevention with students, while a slightly lower figure (47.2%) said they do. This distribution reflects the divide within the profession, which seems to be partly linked to the level of education (greater awareness in junior high school and vocational high schools) and the disciplines (greater awareness in the humanities and social sciences and vocational disciplines).

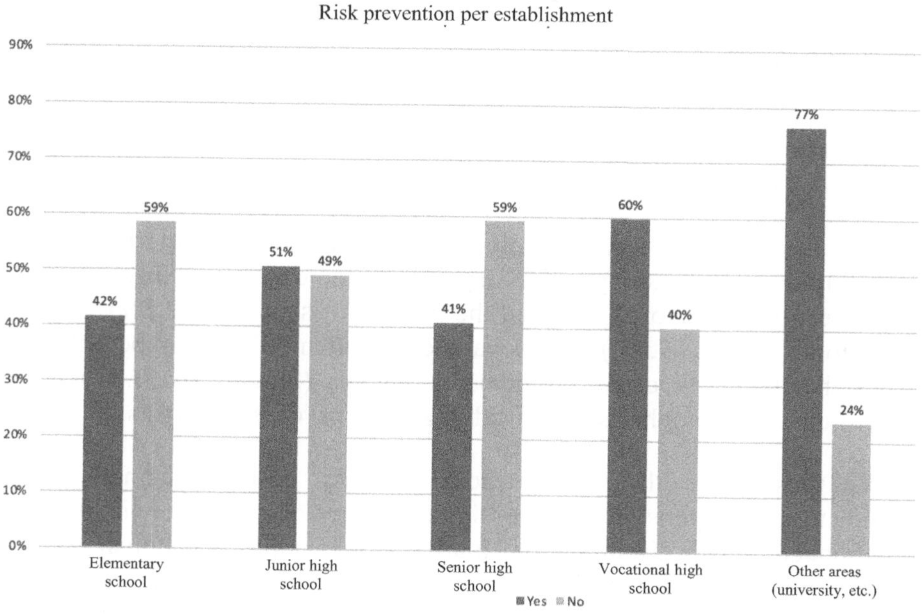

Figure 1.1. *Distribution of respondents addressing digital risks by institution*

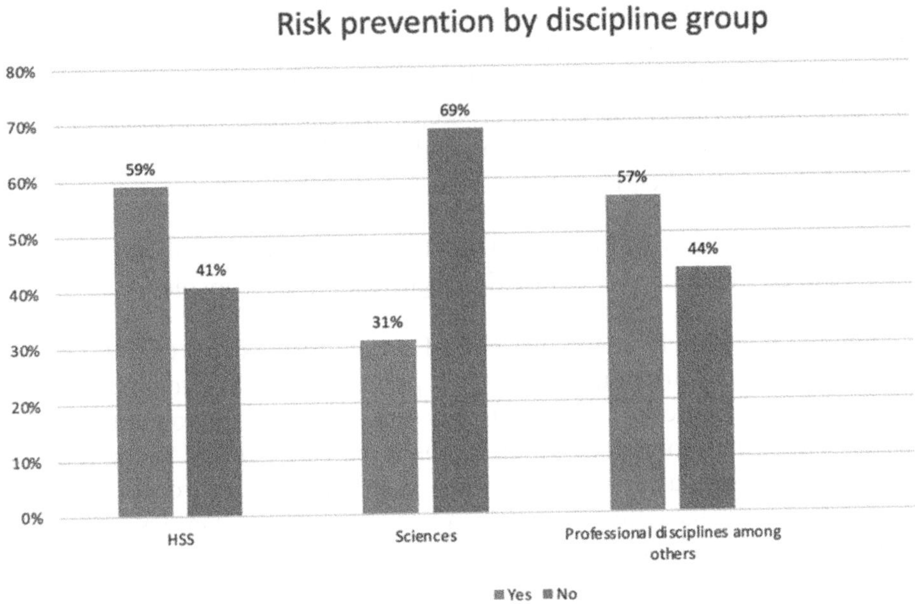

Figure 1.2. *Distribution of respondents addressing digital risks by disciplinary group*

1.4.3. *When risk perceptions lead us to consider them in order to overcome them*

The proportion of teachers who say they consider digital risks when educating students do so by addressing various topics. This mainly concerns the protection of personal data and the privacy of students (89.2%) and the rights to an image (67.3%). This awareness is often linked to cyberbullying issues that regularly cause problems in schools (Stassin 2019). There is therefore a strong emphasis on ethical, legal and psychosocial risks.

We thus notice that some teachers clearly position themselves as legitimate and responsible for the digital literacy of students, without this being linked to their disciplinary affiliation. For example, an English teacher explains that he is involved in the prevention of certain technical and ethical risks by giving students tips on how to choose "strong passwords", thus enabling them to better protect their privacy and data. Others address informational and cognitive risks during Internet searches:

> I tell them "If you go on YouTube, don't click endlessly on the videos on the right, run your search, watch the video that matches your search, then do another search…" (English teacher in a senior high school, age 37).

Among teachers who make daily and varied use of digital technology, their personal experience of the Web can foster pedagogical practices and give them a sense of confidence in helping students secure their informational environment and develop their digital skills. For example, this English teacher says she wants to train students in digital use:

> I discovered computers quite young, I was 8 or 10 years old, so I am used to searching, trying, making mistakes, trying again. And as you develop mechanisms, you learn which symbols correspond to what, you find a particular way of doing something in a section or other of some kind of software. In contrast, I see my colleagues who are twice or three times my age sometimes and they find it much more difficult because they don't have the culture of digital technology that I have been immersed in since I was a child (English teacher in a high school, 25 years old).

This same teacher shows a digital culture that has developed around the various issues related to digital uses. This culture leads her to consider the risks:

> We can limit the use of social networks and control them well, but that doesn't mean we'll relay good information. So, I refer you to the fake news that we talk about so much at the moment, but it's not just related to current events, we can relay completely false information of all kinds, we can be taken in by false ideas, we can even, without going so far, simply read content that is slightly false or politically biased… (English teacher in a high school, age 25).

This teacher is joined by a young high school biology teacher who believes that digital networks are also learning spaces and that young people are not necessarily aware of this:

> So social networks, in my opinion, should not be banned but should be used correctly, [...] there are also advantages to uh…

> Facebook, or Twitter, etc., in that you can monitor information and follow interesting pages. For example, in biology there are very interesting pages. There is the page of the Museum of Natural History which publishes articles, there are the museums of Bordeaux which publish articles, and also American sites. So why not, um, do interdisciplinary work with the English teachers? Explain to the students that ok there are funny things on Facebook but there are also sources of information that are reliable and that they can go and look for information on them and often it's small articles that are not very complicated. And then when it comes to digital literacy, in my opinion it's like all the other types of education, be it sex education, citizenship education, etc. It should be part of the core curriculum. It concerns all the teachers because at the end of the day, our role as teachers is to give students the keys so that they become responsible citizens. And, well, being responsible also means paying attention to what we do with digital technology, and so for me it's something in the core curriculum that all teachers should strive to do, and they don't necessarily do it, so it's a shame (biology teacher in a senior high school, age 22).

These teachers seem to be informed by the media. They are not unaffected by the talk of risk, but they are not overwhelmed by emotions. Instead, they seek to better understand the risks so they can explain them to students. Thus, unlike those who prefer to stay away from all the threats that digital uses can evoke, others, like these teachers, consider that it is essential to address these issues with students and to help young people more generally:

> I'm a teacher, so of course I like to help students and I can't just leave them struggling, but also and above all because very often Facebook and the Internet and social networks are very impersonal spaces where people expose their happiness or their unhappiness without any real human impact in fact and uh… I don't claim to change the Internet, but in any case I use the Internet in a human way and if I see that people are putting up alarmist or sexist, xenophobic, discriminatory messages in general, well, I act. I don't let… I don't let it go. I don't remain indifferent (English teacher in a high school, age 25).

Considering her role as an educator, this teacher tells us that she is even involved in providing mediation and helping young people who are being harassed on social networks. She says that she has joined the association "*Marion 13 ans pour toujours*" and regularly participates in the exchanges that take place on Facebook. She explains that she sometimes offers her help when a teenager appears to be struggling:

> It seemed a bit worrying so I sent her a Facebook message saying, I'm at this association, you can get in touch or you can reply to me and then I'll redirect you somewhere else (English teacher in a high school, age 25).

In the panel of new teachers we interviewed, this category of teachers, who are very committed, is still largely in the minority. Thus, the information and training efforts seem insufficient to go beyond the vision of risks and turn it into a lever for educating students.

1.5. Reflection on the role of digital risk representations in education

When new teachers are asked about their perception of digital risks for themselves and their students, the majority of them see an amplification of risk when it affects students. According to them, students are more exposed than they are to most risks.

This trend can be explained by the idea that teachers have been trained or have trained themselves and feel they have more experience, or because they perceive students as being less armed and feel threatened themselves. This amplification therefore also refers to a sense of personal insecurity with digital technology, as this physics and chemistry teacher shows:

> I get the impression that in the face of social networks and perhaps digital resources in general, there is something that appears, that is to say that there is a little bit of a decline in personal reflection, and again I say this… well, perhaps with some nuance, but I have this feeling. I have the impression that sometimes there is so much information that is given to us, so much ease of access to a lot of information that sometimes… there is a passive attitude that is established in the sense that the

students, but I think it is also true for me and for many of us, we tend to use our personal reflection a little less and that I think is a drawback (physics and chemistry teacher in a senior high school, 30 years old).

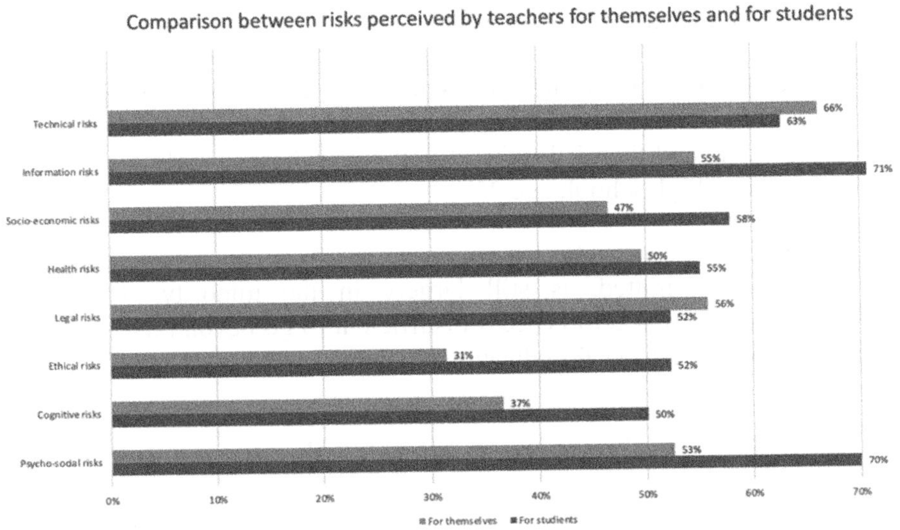

Figure 1.3. *Teachers' perceived risks to themselves and students*

Official texts are also understood in different ways by different people. The level of teaching can have an impact on the representations of what is valued or not valued in teaching. Among those who attempt this venture, there is often a feeling of not being supported by the institution, which gives too few means (equipment, time) to allow the achievement of the requested objectives and imposes too many constraints (school programs, locking certain sites, etc.).

The feeling of belonging to a digital native generation appears to be a lever for teachers who feel better able to transmit and share their know-how.

On the other hand, some of them do not consider digital technology as part of their role as educators and neglect both assistance for its use and the critical approach that underlies digital literacy:

> I know that in history and geography they will do this much more because the subject lends itself to it and then they have the teaching of EMC (moral and civil education) where they… often they will lend themselves to it. It's true that I consider that the students currently handle computer tools much better than I would, insofar as they have access to them [...] But I'm not going to teach them how to use them (math teacher in a senior high school, 31 years old).

In this confusion between technical skills (using the tool) and intellectual and cognitive mastery (understanding/evaluating/criticizing), digital uses do not appear to be an educational issue for these teachers. The (technical or manipulative) skills of the students are not taken into account by the teacher and are even rejected, even though they could be a lever for critical education in digital uses.

The teachers who consider training students to deal with digital risks say that the protection of personal data and privacy on the Internet is the first topic to be addressed (89.2%). Image rights are also cited by 67.3% of them. Next come media literacy and copyright, cited by half the teachers, as well as cyberbullying.

However, only half of respondents (51.8%) report consulting information about digital risks, and more than half of the new teachers who have consulted information on digital risks (54.2%) feel that this information is insufficient. Another 52.3% say they have not been trained in digital uses in teaching situations. Among those who have received training in digital uses, opinions are very divided: half of them (50.1%) state that this training was not useful to them in dealing with the risks; the other half (49.3%) believe that the training enables them to deal with them. The majority of respondents (62.3%) say that training regarding digital risk management with students would be helpful to them. We also note that those who have not been trained are slightly more likely (70.9%) to feel this need. Even among those who have been trained, half (53%) still feel the need for training. The feeling of a need for training is clear, even among those who have already been trained. There is no denial of the need for training, nor is there any dispute about its usefulness.

On the other hand, those who feel that digital risk information is over-represented feel much less of a need for training. In fact, 69.2% of them do not want any particular training. On the other hand, 72.9% of those who consider that information is insufficient request training. Two hypotheses can be put forward to interpret these results. On the one hand, the fact that some teachers have a well-developed information culture may influence their sense of self-efficacy. The fact that these teachers know where to find information gives them the sense that they are able to deal with risks when needed. On the other hand, it can be hypothesized that teachers who feel the weight of media discourses on digital risks instead seek to avoid them. Thus, training on digital risks may be seen by some as counterproductive to practicing digital skills with students.

However, we can see that the more teachers feel that the risks are important for their students, the more they express the need for training. Thus, the nuance lies in the reception of the discourse on risk: while some will consider that the risks are high and require expert skills to deal with them, others will consider that the risk exists but that it remains moderate in their context and that they will be able to manage it.

1.6. Conclusion

The survey of digital risk perceptions among new teachers who are beginning their careers and who are, for the most part, digital natives, reveals their importance and impact on students' digital literacy and the use of digital tools in the classroom. For many, talk of risks has a strong emotional impact that can generate a sense of incompetence and a feeling of lack of training. As a result, these digital natives do not feel more confident or competent in dealing with the issues related to digital uses in their personal context and even more so at school. Alarmist speeches about the dangers of digital technology therefore seem to have a stronger negative impact and do not encourage teachers to engage in educating students through and on digital technology.

Depictions of digital risks therefore often appear to be a barrier to education, especially when they are accompanied by a poorly developed information and digital culture in personal and professional life. In addition to this, there are differences in the idea of their role as teachers and in the interpretation of what the institution expects of them when it comes to training students in digital technology. These very different conceptions testify to the difficulties of the institution and of teacher training in clearly communicating the expectations in this regard. However, we have identified teachers for whom knowledge and consideration of digital risks are a powerful lever for accompanying and educating students in the critical uses of digital technology, provided that they themselves have the skills to become informed and to use different digital tools in their daily lives. Thus, reinforcing teacher training to develop information and digital cultures, particularly by taking into account the diversity of risks, or rather the issues at stake, appears to be a solution in order to respond to the need for critical education on digital technology and its uses in society.

1.7. References

Amadieu, F. and Tricot, A. (2014). *Apprendre avec le numérique : mythes et réalités*. Retz, Paris.

Becchetti-Bizot, C. (2017). Repenser la forme scolaire à l'heure du numérique : vers de nouvelles manières d'apprendre et d'enseigner. Report no. 2017–056, Ministry of National Education, Youth and Sports, France.

Beck, U. (2008). *La société du risque : sur la voie d'une autre modernité*. Flammarion, Paris.

Ben Youssef, A. (2004). Les quatre dimensions de la fracture numérique. *Réseaux*, 127–128(5), 181–209.

Blaya, C. (2013). *Les ados dans le cyberespace : prises de risque et cyberviolence*. De Boeck Superieur, Louvain-la-Neuve.

Bronner, G. (2013). *La démocratie des crédules*. Presses universitaires de France, Paris.

Cardon, D. (2015). *À quoi rêvent les algorithmes : nos vies à l'heure des big data*. Le Seuil, Paris.

Citton, Y. (2014). *Pour une écologie de l'attention*. Le Seuil, Paris.

Cordier, A. (2015). Imaginaire(s) de la jeunesse à l'heure du numérique : entre discours et pratiques, des imaginaires en tension. *Interfaces Numériques*, 4(2), 269–284.

Desmurget, M. (2019). *La fabrique du crétin digital : les dangers des écrans pour nos enfants*. Le Seuil, Paris.

Hayles, N.K. (2016). *Lire et penser en milieux numériques : attention, récits, technogenèse*. Université Grenoble Alpes, Saint-Martin-d'Hères.

Jehel, S. (2015). Les pratiques des jeunes sous la pression des industries du numérique. *Le Journal des psychologues*, (9), 28–33.

Liquète, V. and Le Blanc, B. (2017). Introduction générale. In Les élèves, entre cahiers et claviers, Liquète, V. and Le Blanc, B. (eds). *Hermès La Revue*, 2(78), 11–14.

Merzeau, L. (2013). L'intelligence des traces. *Intellectica – La revue de l'Association pour la Recherche sur les sciences de la Cognition (ARCo)*, 1(59), 115–135.

Musso, P. (2008). La révolution numérique : techniques et mythologies. *La Pensée*, (355), 103–120.

Pariser, E. (2011). *The Filter Bubble: What the Internet is Hiding from You*. Penguin, London.

Plantard, P. (2016). *Les imaginaires numériques en éducation*. Manucius, Paris.

Plantard, P. and Le Mentec, M. (2013). INEDUC : focales sur les inégalités scolaires, de loisirs et de pratiques numériques chez les adolescents. *Terminal*, (113–114), 79–91.

Prensky, M. (2001). Digital natives, digital immigrants part 2: Do they really think differently? *On the Horizon*, 9(5), 1–6.

Rouvroy, A. (2014). Des données sans personne : le fétichisme de la donnée à caractère personnel à l'épreuve de l'idéologie des Big Data. *Étude annuelle du Conseil d'État : le numérique et les droits et libertés fondamentaux*, La Documentation française, Paris.

Stassin, B. (2019). *(Cyber) harcèlement : sortir de la violence, à l'école et sur les écrans*. C&F, Caen.

Stora, M. (2018). *Et si les écrans nous soignaient ? Psychanalyse des jeux vidéo et autres plaisirs numériques*. Erès, Toulouse.

2

Teenagers Faced with "Fake News": Perceptions and the Evaluation of an Epistemic Risk

2.1. Introduction

The phenomenon referred to as fake news is causing great concern in political, journalistic, scientific and educational circles (Huyghe 2018). It revives the problem of disinformation and the risks associated with it. A catch-all word, its use is very common and yet it is challenged and even contested in the academic sphere. Although it is not a completely consolidated scientific concept, the expression fake news has a real-life basis in the media and society, which justifies its use in this study.

The objective of this research is not to propose an objective approach to fake news, but to focus on how it is perceived by a specific segment of the population: teenagers. On the face of it, it could be said that teenagers are particularly concerned by misinformation and disinformation, since they frequently use social media to learn about current events (Aillerie and McNicol 2016), and it is predominantly through these channels that fake news is broadcast (Martens *et al.* 2018).

Nevertheless, while the scientific literature shows that young people have difficulty evaluating whether information online is credible, or whether sources are reputable, in a rational and critical manner (Serres 2012; Sahut 2017), there is currently – to our knowledge – no work in the information

Chapter written by Gilles SAHUT and Sylvie FRANCISCO.

and communication sciences that focuses on their perception of fake news. We have thus opted for a comprehensive approach that centers on the perceptions and practices of teenagers in relation to this informational phenomenon. Our general hypothesis is that this theme constitutes an entry point to better understand their relationship to digital information and their understanding of its epistemic dimension. In other words, their perceptions of the "true", the "false" and the "uncertain" within their informational environment.

This exploratory study will therefore attempt to provide answers to the following questions: What do teenagers believe and know about fake news? Do they feel exposed to a specific informational risk? How do they go about identifying and evaluating this type of information?

To this end, we first propose a synthesis of studies on the production and dissemination processes of fake news and how it is identified and assessed by the public. The aim is to characterize fake news from a scientific point of view to better identify the informational risks it is likely to generate. We then aim to uncover the perceptions held by teenagers on this informational and social phenomenon. Subsequently, we specify the methodology of our study, based on interviews with 14 young people aged 14 to 17. Finally, our results are presented and discussed by comparing them to other studies, particularly those concerning credibility judgments. They will thus be considered in relation to questions on Media and Information Literacy (MIL) and, more particularly, through reflections on the teaching practices used to enable young people to critically evaluate information.

2.2. Fake news: from production to reception

Since 2016, the topic of fake news has given rise to a large number of studies combining computer science and the social sciences. Most of this research focuses on the process of its creation and propagation, the actors involved, their motivations and the detection systems developed in an attempt to deal with it.

We do not claim to have summarized the hundreds of articles published on this subject (for a synthesis, see Kumar and Shah 2018; Zannettou *et al.* 2019). We simply highlight the characteristic features of fake news, which leads us to identify the potential risks it generates. We then establish that there are far fewer studies on how the public perceives the credibility of fake news; however, they are essential in order to understand the real risks involved.

2.2.1. *Characterizing the fake news phenomenon*

While use of the term "fake news" has become commonplace since the 2016 US presidential campaign, it retains a certain semantic vagueness. Researchers themselves are divided on the use of this concept in an academic setting, and when they do use it, it is clear that it encompasses various realities (Tandoc *et al.* 2018). Without purporting to provide a universally accepted definition, we can discern traits that characterize this informational phenomenon:

– fake news concerns current events: politics, of course, as well as news, health (vaccination, nutrition), finance, etc. (Lazer *et al.* 2018; Zannettou *et al.* 2019);

– it is spread via social media with the aim of reaching a wide audience (Allcott and Gentzkow 2017);

– it imitates journalistic forms and codes: articles, captioned press photos, live streams, etc. (Starbird 2019);

– the information is not only *false*, but fabricated and falsified (*fake*) to appear credible and thus deliberately mislead the reader or viewer.

Defined in this way, fake news constitutes a particular category of the broader concept of disinformation, understood as inaccurate information intentionally issued by a source seeking to create false beliefs (Fallis 2015). It can therefore be distinguished from the concept of misinformation, which can be defined as inaccurate information issued by a source that is unaware of the erroneous nature of its statement (Søe 2016).

Fake news producers may pursue one or more political or ideological goals: they seek to support a worldview, sow confusion, or weaken an opponent (Huyghe 2018). They may also be motivated by commercial interests: generating traffic on websites to sell advertising space. Fake news then takes the form of *clickbait*, which sports deliberately misleading titles to attract the attention of Internet users and encourage them to click on a hypertext link; the latter is then exposed to advertisements that generate revenue for the source. Finally, some individuals – sometimes referred to as trolls – disseminate fake news for entertainment purposes or to gain visibility and popularity on the Web (Zannettou *et al.* 2019).

To understand the finer details of the fake news phenomenon, it must be repositioned within a more global information and communication framework. While document falsification has a long history, the digital ecosystem provides particularly fertile ground for its propagation (Latzko-Thot 2018). The production and dissemination of fake news are indeed facilitated by:

– the "modifiability" of the digital document: word processing, image editing and audiovisual editing software are now in common use. They encourage manipulation, in the sense of data processing, as well as of maneuvers designed to distort reality;

– the multimedia dimension of digital technology: the ability to combine documents from different semiotic systems (written, still or moving images, sound) makes it possible to produce "documentary evidence" to support a false or imagined assertion (Rebillard 2017);

– self-publishing on the Web: this technical possibility makes it possible to do without the editorial filters inherent in print culture;

– the existence of online communities based on social mechanisms (forums, digital social networks, collaborative sites): fake news is thus developed within the framework of exchanges on forums and then disseminated via social media (Zannettou *et al.* 2017);

– the speed of information dissemination: the interconnections inherent in digital social networks favor the virality of false information, which spreads more quickly and widely than proven information (Vosoughi *et al.* 2018);

– automation of dissemination: software called botnets, linked to fake accounts on social media, contribute to the massive propagation of fake news (Zannettou *et al.* 2019);

– the international dimension of communication on the Web: governments of authoritarian countries – as well as of democracies – employ "cyber soldiers" for the purpose of manipulating information, both for their own population and those of foreign countries (Bradshaw and Howard 2017). Note that the "cyber soldiers" themselves may be financially motivated, like the students in Macedonia who produced a massive amount of fake news during the 2016 American elections and made substantial profits from it (Mercier 2018).

2.2.2. The potential risks associated with fake news

The concern that surrounds fake news stems from the risks that it can generate at both an individual and collective level. This is a particular type of informational risk that we refer to as epistemic because it is linked to the truth value of information. At the individual level, being misled by fake news is, at the very least, a distraction and waste of time; the consequences can be more serious, however, if the deception is not identified. Indeed, believing in false information can bias acquired knowledge, as well as lead to harmful decisions and actions. One need only imagine the potential harmful effects on health, a field particularly affected by the disinformation that circulates on digital social networks (Waszak *et al.* 2018). It is also important to acknowledge the tension that is created by the uncertainty of the truth value of information. While too much credulity can be problematic for the reasons mentioned above, the increased fear of being duped can lead to an exacerbated distrust of sources and to depriving oneself of potentially helpful information.

The informational risks associated with fake news may also have a societal dimension. Indeed, fake news contributes to the counter-discourse that fuels distrust of traditional authority figures, such as journalists, politicians and scientists (Badouard 2017; Proulx 2018). Fake news is not only likely to deepen this distrust, but also to generate confusion and the feeling that it is impossible to access true facts. The possible consequences are, on the one hand, disinvestment in political life and, on the other hand, the impossibility of a democratic debate. The latter is based on shared recognition of proven facts, which constitute common references from which interpretations and arguments can be exchanged. The term post-truth thus reflects the concern that emotions and personal beliefs will take precedence over the consideration of objective facts and rational arguments (Mercier 2018; Revault D'Allonnes 2018).

In a similar way, the issue of fake news is linked to that of echo chambers and filter bubbles, phenomena in which individuals are exposed to homogeneous content according to their ideological preference (Bakir and McStay 2018). Empirical studies have shown that social and political homogeneity is a driving force behind the spread of disinformation on digital social networks (Del Vicario 2016; Starbird 2017). This homogeneity thus serves to polarize opinion and fragment the public space. In this way, we can see the threats fake news poses to democratic life. By fueling epistemic uncertainty and exacerbating divisions, it constitutes an obstacle to the ideal of collective deliberation based on rational reasoning.

However, the severity of the risks mentioned here remains a matter of debate (Cardon 2019). It requires an understanding of the public's level of exposure to fake news, as well as their attitudes towards it. A study carried out in the context of French information provides interesting insights into these points (Flechter *et al.* 2018). The study shows that sites considered to be vectors of fake news attract a relatively low monthly audience (between 3.1% and 0.2%) compared to traditional media sites (22.9% for lefigaro.fr, 19% for lemonde.fr). But it also highlights that the fake news on these sites generated as many or more interactions ("likes", shares or comments) on Facebook than the information coming from traditional media. These interactions demonstrate an interest in fake news, but do not prove that the content is endorsed. The individuals who share or comment on them may simply be trying to point out the surprising or spectacular nature of the information, or even trying to denounce a manipulative intention. In order to estimate the extent of the epistemic risks created by fake news, it seems necessary to look at public attitudes towards it, particularly the credibility they attach to it.

2.2.3. *The credibility of fake news*

A significant amount of research has been done on the processes for assessing the credibility of information, the degree of competence of different audiences and the possible influence of different variables (age, gender, level of education, socio-cultural background, etc.) on these processes. It is impossible for us to propose a synthesis of these works here. We therefore limit ourselves to pointing out just two trends that seem to be emerging among young people: on the one hand, their tendency to favor criteria linked to the usefulness of the source, and its ease of access and use (pragmatic judgments), over criteria for judging the credibility of information and the reliability of the source (epistemic judgments); on the other hand, the fact that these epistemic judgments are based more on heuristic processing (quick, intuitive, surface-level evaluations) than on an analytical process involving reasoning that is based on multiple criteria and greater interaction with the source to be analyzed and the information it presents (for a synthesis, see Sahut 2017).

Work that focuses exclusively on the evaluation of fake news is limited and only very recent. In psychology, attention has been given to the role played by a specific heuristic: confirmation bias in the credibility of political fake news. Craig Harper and Thom Baguley's (2019) study of a sample of American liberals and conservatives, as well as English Brexit supporters and opponents, indicates that politically engaged individuals are more likely to believe (at least partially) stories that are consistent with their partisan beliefs. Similarly, they tend to doubt the credibility of real news stories that contradict their views. Conversely, a study by Gordon Pennycook and David Rand (2019) emphasizes that it is not ideological preference that determines the credibility of fake news, but a cognitive variable. According to them, individuals who favor analytical reasoning processes (i.e. non-heuristic) are more capable of discerning this type of disinformation.

Within the information sciences, we identified two research studies related to the credibility of fake news. One study surveyed 2,747 American adolescents between the ages of 11 and 18 to assess their ability to identify hoax websites (Metzger *et al.* 2015). The results showed that more than half of the young people surveyed said they believed – at least partially – information from these sites. Older teens showed greater distrust of these hoax sites; they were more aware of issues related to a possible lack of validity of online information and reported using more analytical strategies for assessing credibility.

The second study involved a group of 63 American undergraduate students (Leeder 2019). Participants were asked to assess the credibility of a sample of articles – of which half were fake news stories – and explain their approach. In the end, students correctly identified 64% of the fake news articles and 60% of the real news stories, indicating heterogeneity in the mastery of this informational skill. Those who performed better spent more time assessing articles and examined the entire webpage to judge the credibility of the information. Thus, the findings align with Metzger *et al.* (2015) and Pennycook and Rand (2019) in terms of the importance of analytic processes as a means of identifying fake news.

To complete these quantitative studies on the informational skills of young people, we opted for a qualitative piece of research to gather their individual and collective views on fake news. It seemed essential to us to understand how teenagers express, in their own words, their perception of this informational phenomenon, how much importance they attach to it and the meaning they give to it. This comprehensive approach seemed to us to be all the more appropriate as fake news is the subject of social discourse likely to feed the views of teenagers.

In the context of information and communication science, research has highlighted the weight of young people's perceptions of the Internet (Cordier 2011) and Wikipedia (Sahut 2014). Often far removed from scholarly knowledge, these perceptions have an influence on information practices. For example, Wikipedia's largely negative reputation within academia changes the perception that high school and university students have of the benefit–cost ratio linked to consulting this source. It introduces the notion of risk: the risk of being confronted with the "false" and the risk of being sanctioned by teachers who are against the use of Wikipedia if this source is cited in schoolwork (Sahut 2014). The question is therefore whether young people have been exposed to distrust of fake news and whether it has affected their epistemic trust in digital sources.

2.3. Methodological framework of the study

In order to collect the data, we conducted 14 individual, semi-structured interviews with young people aged 15 to 17 years old who were enrolled at a general education high school during 2018. To encourage the interviewees to express themselves as freely as possible, we developed an interview guide that alternated between phases in which they would be questioned (directive phase) and those in which they could express themselves without being questioned (free expression phase). Care was taken to ask questions that were sufficiently open-ended to avoid conditioning the response, in accordance with the recommendations of Blanchet and Gotman (2015). The questions asked of the 14 young people in our sample were aimed at determining their views and their information practices related to fake news: Have they heard of it before? What definition and examples do they give? Do they feel they are exposed to it? Do they feel equipped to deal with it? How do they go about identifying it?

In order to limit the social desirability bias that such a subject may generate, we guaranteed the anonymity of the young people interviewed and the absence of value judgments on their words. For the same reason, the interviews took place at their homes or by telephone and not in a school setting, which could have influenced their responses. Finally, inspired by a methodological reflection developed by Vermersch (1994), we frequently asked the teenagers to share examples about fake news from their own personal experience. In this way, we expected the interviewee to draw upon their episodic memory and not their semantic memory, which limits the use of stereotypes and platitudes assumed to be in line with the interviewer's expectations. During the interviews, we tried to adopt a benevolent attitude, both verbally (various forms of approval) and physically (smiles, encouraging looks). Whenever appropriate, we rephrased the interviewee's comments to ensure that we had understood them correctly, reopened the discussion without asking new questions or reassured the interviewee that we were listening.

Name	Age	Education level	Duration of the interview	Location of the interview
Manuel	15 years old	Ninth grade	7'	At home
Ambre	15 years old	Ninth grade	19'	At home
Anna	15 years old	Ninth grade	15'	At home
José	16 years old	Ninth grade	10'	At home
Christian	15 years old	Tenth grade	15'	At home
Sophie	15 years old	Tenth grade	17'	At home
Yves	15 years old	Tenth grade	17'	By phone
Maco	15 years old	Tenth grade	36'	At home
Nassim	15 years old	Tenth grade	26'	By phone
Lisa	16 years old	Tenth grade	11'	By phone
Thibault	16 years old	Twelfth grade	20'	At home
David	17 years old	Twelfth grade	33'	At home
Sylviane	17 years old	Twelfth grade	19'	By phone
Iris	17 years old	Twelfth grade	17'	By phone

Table 2.1. *Summary table of interviews*

After transcription, the comments collected were subject to a cross-cutting thematic analysis. This allowed us to compare the responses of the 14 young people in our sample and identify common traits and differences.

The highly variable duration of the interviews already appears to be an element that can be interpreted. It may of course reflect the personality of the respondent (chatty or shy), but if we cross reference this with the content of their remarks, we note important differences in their degree of reflexivity. Some of them (Manuel, José, Lisa) had real difficulties in "putting into words" their informational experience, while others (Sophie, Sylviane, Iris) showed that they were capable of having a structured conversation about their practices and revealed that they had given substantial thought to the issues related to fake news. We also asked the young people about their use of digital social networks and noticed significant disparities in their level of participation. Some, like Christian or Lisa, did not have a strong presence on these networks, while others use them assiduously.

	Snapchat	Instagram	Messenger	WhatsApp	Facebook	Twitter
Manuel	•	•	•		Δ	
Ambre		•		•		•
Anna	•	•		•		
José		•			Δ	
Christian					Δ	
Sophie	•	•				
Yves	•	•				•
Maco				(Δ)		
Nassim	•	•	•		Δ	
Lisa				(Δ)		
Thibault	•	•			Δ	•
David	•	(Δ)			Δ	
Sylviane	•		Δ			
Iris		•	•	•	(Δ)	•

Table 2.2. *Interviewee presence on social networks •: Active account; (Δ): Account not very active; Δ: Inactive account*

2.4. Results of the study

We first present our respondents' views on fake news (understanding of the term, the intentions of those producing it) and the sources of knowledge of this phenomenon. Then, we report on the processes for identifying fake news and critically evaluating the sources at work, as well as perceptions of the severity of this epistemic risk.

2.4.1. *A heterogeneous understanding of the concept*

The English expression "fake news" was known to some of the young French people interviewed. They gave a definition close to the notion of disinformation, including the idea of a manipulative intention.

> Well, basically, it's information that's fake and that's made to make people believe it's true, so to trick people, essentially (Anna).

> It's when it's something fake. When it's not true. When it's a lie (Christian).

Others were more hesitant and gave more vague definitions, omitting any mention of the intentional nature of fake news.

> I've heard that word before, on TV, I think. But then, I haven't really looked up in detail what it is (David).

> Fake news? Yeah, I think, that sounds familiar… Yeah. Fake news? As in, fake stuff? (Lisa).

When asked about examples they encountered in their daily lives, some of the young people interviewed mentioned one or two specific instances where they were aware of having been confronted with information that turned out to be false.

> XXX Tentacion is a rapper who died and there had been fake news circulating that he wasn't dead, that it wasn't true, that it was to promote his new album, when in fact he was really dead. That's when I fell into fake news (Anna).

> Fake news tends to be about gossip and celebrities, more than anything else. After we won the Cup, they said that Nabilla and M'Bappé were together, and that really blew up (Iris).

Rumors are thus considered fake news (eight mentions), whereas in the examples given, the deliberate intention to deceive – a characteristic feature of fake news – is not obvious[1]. The notion of fake news is also associated with other information pollution, such as advertisements (three mentions), spam (three mentions) and commercial scams (five mentions):

> On social networks, it's mostly ads, spam, things that don't necessarily interest us. There can be scams like we see sometimes in email. Often, there are also things for dating sites. In general, that's it (David).

The following was also considered to be fake news:

– enticing links, qualified by young people as "clickbait" (nine mentions): "It's when there's something marked in the title but in the video, there's nothing like that. It's totally wrong, actually. It's 'clickbait'" (Jose);

– phishing or hacking (two mentions): "Windows that open and say, 'You have 5 viruses on your computer: click on this link to solve the problem'" (Christian);

– the use of the term fake news for political rhetoric (two mentions): "Trump who calls all information fake news when it doesn't suit him..." (Sylviane);

– political lies (one mention): "I remember that when Chernobyl happened, the authorities said that the radioactive cloud had not passed over France. They lied. To keep the population under control, I think" (Christian);

– urban legends (one mention) (the case of the Momo Challenge is developed below);

– trolling (one mention): "False polemics on events like the attacks" (Nassim);

– media exaggeration (one mention): "When they say a bank was robbed when it was a grocery store" (Nassim).

---

1 Rumor can be defined as "the dissemination of unverified, functionally relevant information statements that appear in contexts of ambiguity, danger, or potential threat and help manage risk and understanding" (Di Fonzo and Bordia 2006:23). Not all rumors are fake news. According to the definition we have adopted, only those that turn out to be false, are deliberately launched and widely disseminated can be considered as such.

For teenagers, the term fake news seems both polysemous and all-encompassing. It is used to designate informational risks related to "fakery" and deception. There is some confusion between the different types of digital risks. Sometimes, the notion of fake news is fused with other concepts with which it has only vague connections.

2.4.2. *A blurred perception of the goals of fake news*

The teenagers interviewed do not always identify the motives of fake news distributors, or their possible sociological identity; when they do, it is the commercial goals that are highlighted (nine mentions). This idea is linked to experiences on social networks, notably with "clickbait". The fact that certain information has very catchy headlines, designed to capture their attention and elicit action on their part, was highlighted (four mentions). The idea of virality associated with a quest for an audience is apparent in certain comments. This includes "buzz" and popularity linked to the number of "views", "likes" and "retweets". However, the advertising mechanisms based on the activation of hypertext links (click economy), a real financial windfall for the creators of fake news, were hardly mentioned. This ignorance leads them to imagine the economic processes at work:

> I think people who make fake news are looking for buzz on Twitter to be popular. Getting "likes" and having their posts retweeted because it makes them money. I don't know how but I know that when you retweet, they make money (Yves).

Sylviane is an exception: her reasoning led her to make the link between seeking an audience, exposure to advertising and economic revenue:

> It's known as 'clickbait'... I don't have a precise explanation, but I think they probably earn money every time people visit their site, through ads or whatever. So, their goal is to have as many people as possible visiting their site. No one has explained it to me clearly, but that seems logical enough (Sylviane).

The political or ideological motivations behind fake news (four mentions) are clearly perceived much less than the commercial motives, although they are widely used in the media and in political and educational discourse on the subject. The idea of influence and psychological manipulation was occasionally mentioned but without any connection to politics.

> It influences a lot of people. It's a mass influence (Maco).

> A young person can go on Facebook more than 10 times in a day, meaning that they might come across the same information several times. So, it will get into their head. They might click on it, look at what it is. And if they're not careful, they may come across false information. It might lead them astray (David).

2.4.3. *The diversity of fake news sources*

Fake news becomes known through the browsing experience and/or through preventive rhetoric. Some teenagers admit to having been tricked and having clicked on sensationalist headlines (six mentions):

> Sometimes I click on it and read because there tend to be questions asked in the title. You have to click on it to see the whole article and get the answer (Nassim).

> I've probably been tricked before I was told to watch out for fake news (Thibault).

Some of them were also exposed to preventive messages. A minority (four mentions) reported being educated on the subject in a school setting, either in history, geography or economic and social sciences classes, or by a librarian. The reception of these pedagogical interventions proved to be extremely variable. Anna recalls becoming increasingly aware of it during her seventh grade year: "Before the course [run by the librarian] I couldn't conceive that there were false things on the Internet and since then, I told myself that everything that was said there was not necessarily true". Iris, on the other hand, puts the contribution of the intervention that took place in the context of history and geography into perspective: "We learned some interesting things but afterwards... Yep... When I left the course, I didn't say to myself 'this is going to change my life'. Maybe it wasn't useful, but it was interesting. It was about prevention, but we already know that". We see here that the perceived usefulness of preventive discourse differs depending on the respondent's knowledge and command of this informational issue.

Awareness of the phenomenon of fake news can also take place in a non-formal setting. Some of the young people interviewed said that they had received warnings from their parents about the unreliability of the information available on the Internet:

> Parents and the people around us also tell us that we have to be careful with what we see [...]. Well, my parents didn't take an hour to explain it to me but, sometimes, they tell me little things like that… When I show them things, they say, 'Well, be careful, that site looks fake to me' (Sophie).

> My father tells me, 'Be careful, it's fake news'. He tells me, 'Be careful, with social networks and all that, you always have to check' (Iris).

The role of the family environment in the digital risk prevention process is thus brought to the fore (Cordier 2015). However, on this topic specifically, parental warnings do not seem to be widespread as less than half of the respondents mentioned them (six mentions). In addition, several young people in our sample referred to YouTubers who raise awareness around the risks associated with disinformation (five mentions):

> There are YouTubers who warn against fake accounts of theirs. Cyprien, for example (Thibault).

> On YouTube, some well-known YouTubers know that people can misappropriate what they say. So, they sometimes make videos to warn people. They do prevention. I remember Cyprien did something like that… (Maco).

On this subject, Ambre talks about a particular experience. She was initially alerted via a video by Sora – a YouTube gamer – about the dangers linked to Momo, a harmful character who haunts the Web: "Sora said that if you wrote a message to Momo, he could collect data on you and blackmail you, push you to commit suicide or do things, you know". Ambre then set out to find more information on the subject. On WhatsApp, she found a conversation between an Internet user and Momo, which she interpreted as proof the character exists. Sometime later, she said she learned that Momo was, in her words, "fake news"[2]. And in this case, it was the YouTuber's warning broadened its reach. We also see that Ambre implements a real strategy to verify the information on the subject by consulting a variety of sources.

It should also be noted that some respondents mentioned other sources that contribute to fake representations and knowledge: television (three mentions), Twitter (one mention), discussion forums (one mention) and even a streaming series, Quantico, which features the political use of fake news (one mention). Fact-checking sites were only mentioned by Iris. None of the respondents mentioned fake news as a topic of conversation among peers.

2.4.4. *Identifying fake news: heuristic processing and analytical strategies*

Several teenagers interviewed mentioned the ways in which they recognize fake news and the means they use to verify its credibility. Judgmental heuristic processing associated with a sense of distrust was mentioned. For Iris, it is the spelling errors that act as a warning signal: "When there is a site where there are spelling mistakes, I am a bit suspicious". Anna mentions another type of heuristic processing:

> Actually, it's strange when something might be fake news, when you open the link, there's something weird. Maybe it's the way it's presented. For example, when there are tabs on the side with small writing, with a certain color, it can be a bit odd… it alerts me a little bit (Anna).

Anna's comments reflect her browsing experience which has enabled her to identify a layout – and elements related to graphics, colors and typography – that she considers typical of fake news. Her epistemic judgment, which is based on the esthetics of the site and not on its informational content, bears witness to the implementation of a heuristic process based on visual appearance (Hillgoss and Rieh 2008). Similarly, "headlines" (David) that

---

2 Momo is more like an urban legend and a hoax whose origin has not yet been determined, to our knowledge at least. Saferinternet.at (2019). Achtung HOAX! Gruselige Nachrichten von Momo [Online]. Available at: https://www.saferinternet.at/news-detail/achtung-hoax-gruselige-nachrichten-auf-whatsapp-von-momo/ [Accessed 17 February 2019].

are "catchy" (José) or "hyper-catchy" (Sylvaine) were mentioned as possible indicators of fake news.

On digital social networks, some rely on comments as credibility cues (two mentions):

> On Twitter, at first, I kind of believe it but then I go and look at the comments. And I see that often, there are lots of people who mark "fake news", that it's false and everything. So, then I don't believe it anymore. I mean, if there's only one message that says it, I'll still believe it, but when I see that there are many… (Yves).

A social heuristic process known as recommendation (Metzger *et al.* 2010) is highlighted here: Yves takes into account the judgments of others to construct his own evaluation.

On Twitter, specifically, two respondents refer to another credibility index, account certification, which they say is supposed to protect them from fake news:

> I'm looking at the blue thing. It says it's official, so it's safe. [Meaning?] Well, that means it's reliable (Ambre).

> With the certified accounts, I have confidence. Because I know that if there are false things, well, someone will… they'll get a fine, or they'll shut them down. Certified accounts are sure to be legitimate, you know… But if they're not certified accounts, I'm suspicious. Because, sometimes, the information is weird. And, on top of that, I see that it's not certified information. So that makes me even more suspicious (Yves).

What Ambre and Yves refer to can be described as an expertise heuristic process (Sahut 2017). They are indeed looking for a single clue that reassures them of the reliability of the source and thus the validity of the information. However, the blue certification badge on Twitter is primarily intended to fight against impersonation. Granted upon request, it only guarantees the identity of the source and not the truth value of the information, as Twitter is not a validation authority. The certification of the account is withdrawn in specific cases, such as those involving hate speech or harassment, but this is not the case when it comes to spreading fake news[3]. For Yves, his lack of knowledge about this mechanism causes him to resort to his imagination and leads him to erroneous beliefs about Twitter's editorial model.

We also identified analytical approaches to critically evaluate sources that differ from the heuristic processing mentioned above. First, teenagers reported more sophisticated strategies that may involve extensive content analysis or consideration of cues that indicate the reliability of the source (four mentions):

> I try to sort through what I'm looking for. I go and look at some of the sites that are out there. I read the content and try to get an idea. I get suspicious if I see things that don't make sense (Maco).
>
> If I do a Google search, I look carefully at the results. I try to see who the publisher is, as my ES (economic sciences) teacher explained to me. And for photos, I look at the date, like my history and geography teacher told me (Christian).

Second, most of the young people interviewed said that they verify information they are not sure about (eight mentions). Some made use of online research as a means of corroborating sources:

> I'll check if other sites are talking about it and which sites are talking about it. If it's gossip sites or Le Monde. (Sophie)

Others say they check the information they doubt with adults around them whom they consider competent, most often their parents (four mentions):

> Depending on what I want to check, I choose the person who seems most competent. It can be my parents, my teachers, a specialist in the subject… (Christian).
>
> The time I read that the European Cup was going to be replayed because the referee did not do his job properly, I thought to myself: it's a strange thing for them to say. But I half believed

3 Twitter (2021). FAQ sur les comptes certifiés [Online]. Available at: https://help.twitter.com/fr/managing-your-account/twitter-verified-accounts [Accessed 20 September 2019].

> it. I went to ask my parents. They confirmed that it was wrong (Sophie).

Ultimately, they allude to a reassuring information environment designed to protect them from epistemic informational risks.

2.4.5. *A remote and controlled phenomenon?*

The teenagers interviewed have different understandings of the significance of fake news and the issues relating to it. Several of them showed indifference to the phenomenon: either they do not feel concerned: "I don't really care. I've hardly ever heard of it. It doesn't particularly concern me" (Manuel), or they feel like they are able to avoid the risks by controlling their information environment: "I'm not too exposed. I'm very careful. So…" (Christian).

Almost all of them say they are very careful about the sources they use, only going to those that are: "known and safe" (two mentions), "trustworthy" (three mentions) or "super reliable" (one mention), "official" (three mentions), "classic" (one mention), "reference" (one mention), "secure" (one mention), "specialized" (one mention). This is to avoid being "scammed" (three mentions), "spied on" (one mention) or "infected" (one mention). When asked to cite sources they consider reliable, they referred to: ONISEP (one mention), Parcours Sup (one mention), Le Monde (six mentions), France Inter (two mentions), BFM TV (3 mentions), L'Équipe (two mentions), France 2 and France 3 (two mentions), Arte (one mention), CNews (two mentions) and Usbek & Rica (one mention). Half of the young people interviewed also said that they get their news from cell phone applications (e.g. upday, News, Discover), which, in their view, protects them from disinformation as the information comes from recognized media.

Only a few of the teenagers admitted to feeling particularly concerned about fake news, most notably when they deemed it to be high stakes (six mentions). In other words, when the information was about a topic of importance, such as terrorist attacks, or which directly concerned them (high school reform, for example): "If it talks about something in France or which is going to change things, or if it talks about, I don't know, the baccalaureate, the new reforms, yes, I'm really going to want to know if it's true or false" (Sophie). However, the societal and political issues surrounding fake news were hardly mentioned. The vast majority of the young people interviewed do not seem to be aware that disinformation could pose for democracy.

2.5. Discussion of the results and reflections on media and information literacy

This exploratory study has demonstrated that a disconnect exists between media, political and educational discourse on fake news and the views of teenagers. The former emphasize the danger that fake news poses to democracy, whereas for the majority of the young people interviewed, it represents a limited, controlled and, for some, even non-existent risk. This discrepancy is partly due to differences in how the concept of fake news is perceived, the same term being used to identify different realities.

Our study shows that, with some exceptions, teenagers' knowledge of this information phenomenon remains superficial and imprecise with respect to the academic definition we have proposed[4]. The term fake news is associated with a heterogeneous set of digital risks: epistemic risks linked to disinformation, as well as technical risks (viruses, hacking), financial scams and unwanted exposure to advertising. The comprehensive approach adopted has therefore allowed us to bring to light a mixture of problems that teenagers actually encounter in their digital practices and the work of the imagination, which attempts to understand, or even explain, these problems. At the same time, these findings lead us to emphasize the value of establishing a conceptual model for the teaching of digital and information literacy (Serres 2007). It seems to us that students' understanding of a global typology of digital risks – including issues related to self-exposure on digital social networks and cyberstalking – could foster more precise identification of the nature of the various problems and allow for a more discerning look at the appropriate means of protecting oneself. In the same way, the current emphasis on fake news should not make us forget the value of raising awareness about other types of disinformation (propaganda, political and/or media lies, etc.) and misinformation.

4 Nevertheless, it should be remembered that the academic sphere is itself confronted with definitional problems on this subject. But some of the young people interviewed give much broader definitions than those circulating in scientific circles.

We found that the motivations of the creators of fake news were largely ignored by our respondents. The economic model at work, based on capture of the Internet user's attention, exposure to advertising and the economy of the click, is sometimes suspected, but more often than not, only superficially understood. This makes the case for media and information literacy that exposes the economic logic of information production in digital industries, often invisible to the user (Jehel and Saemmer 2017). The lack of in-depth knowledge of the possible ideological and political effects of disinformation also challenges the intended goals and implementation modalities of MIL. On the one hand, in the context of information communications, the reading and analysis of a document is strongly influenced by the reader's knowledge of the information producer and the associated expertise and communicative intentions (Tricot *et al.* 2016). In this sense, the critical reader is one who identifies the effects sought by the source – particularly from an ideological point of view – and knows how to distance themselves from it. On the other hand, even if the question of the possible effects of fake news – and beyond that, of disinformation – on the public is far from being clear-cut, addressing this problem with students could encourage reflexivity on a civic issue that is frequently on the political and media agenda[5].

The perception of the phenomenon is often shaped by warnings that emanate from institutional cognitive authorities, such as parents and teachers, or from new incarnations of authority, such as YouTubers. But the concept of fake news is ultimately reinterpreted by each individual and takes on meaning from their respective informational experience. Thus, if the epistemic risks seem minor, it is because in most cases the phenomenon has been encountered in the sphere of leisure activities where the effects of the "fake" may seem "annoying" but not actually harmful. When it concerns fields where the informational issues are perceived to be more important, the concern becomes stronger.

The perception of epistemic risk is also mitigated by the feeling that teenagers have of controlling their informational universe and thus considerably reducing the uncertainty resulting from the new media landscape. Even if information behavior via cell phone applications seems,

5 Recall, for example, that the law on combating false information has been hotly debated. See, for example, Couronne, V. (2018). Loi anti-fake news : une fausse bonne idée. INA : *La revue des médias* [Online]. Available at: https://larevuedesmedias.ina.fr/loi-anti-fake-news-une-fausse-bonne-idee [Accessed 5 January 2021].

on the face of it, to keep some of them away from fake news, this juvenile comment must be put to one side. On the one hand, we know that young people tend to overestimate their informational skills, especially those who lack them the most (Flanagin and Metzger 2010; Gross and Latham 2012). On the other hand, we are aware that there can be a third-person effect in persuasive communication: a person exposed to persuasive messages generally views them as having a stronger effect on others than on themselves (Davidson 1983) and therefore tends to underestimate the actual epistemic risks that they may be exposed to (Jang and Kim 2018). Moreover, the teenagers interviewed may have encountered forms of disinformation without being aware of it.

Beyond the issue of fake news, this study provides an insight into views of epistemic trust, as well as the evaluative practices implemented. The teenagers interviewed repeatedly assured us that they refer to "official" and "reliable" sources, to traditional media (from Le Monde to BFMTV) that are supposed to guarantee the accuracy of the information. These are presented as editorial entities whose credibility is self-evident and who are considered reassuring in a context of informational uncertainty. However, there is a virtual absence of arguments on the basis of this epistemic trust. Thus, within the framework of media and information literacy, aimed at formulating more refined and critical assessments, one of the objectives could be to enable students to more clearly distinguish between different types of institutional cognitive authorities: political and administrative institutions, media institutions, knowledge institutions linked to research and education and, above all, to identify both the epistemic norms that govern them and the factors (economic, material, temporal, etc.) that mean that these norms are not always respected.

Also in the area of epistemic judgments, our respondents reported judgment heuristic processing (visual or aesthetic, social, expertise) that, according to them, favored the identification of fake news, but they also reported analytical strategies (analysis of the plausibility of the content, the identity and expertise of the source, corroboration) that testify to efforts made to evaluate the truth value of the information. Our study, which is based on participants' statements, does not allow us to know the respective frequency of these two modes of evaluation. Recall here that the literature on the subject indicates that the use of heuristic processing is more frequent than the use of analytic strategies (Sahut 2017). Yet, on this point, studies are convergent: analytical approaches are more effective in identifying fake news (Metzger *et al.* 2015; Leeder 2019; Pennycook and Rand 2019). But the costs of learning and implementing these approaches remain higher than those associated with the acquisition of heuristics, spontaneous and intuitive modes of operation. Cognizant that learning to critically evaluate information is complex and requires pedagogical progression that is in line with teenagers' cognitive development (Metzger *et al.* 2015), the following questions can be asked: Within the framework of MIL, is it conceivable– during the first years of junior high school, for example – to aim for the acquisition of heuristics (expertise heuristics or even visual heuristics, insofar as fake news can appear in similar forms, as respondents have underlined), in order to show the limits of these heuristics and to teach real analytical strategies? Or should we teach an analytical evaluation process first, the risk being that this approach is out of reach for the youngest?

2.6. Conclusion

To conclude, we point out the limitations of this study, which present as many potential avenues for future research. This study is primarily exploratory. It only involved 14 teenagers, and we cannot claim here that the sample is representative of the entire age group. For example, we did not interview teenagers who were enrolled in vocational high schools or those who had dropped out of school. Despite this, the interviews conducted revealed significantly different levels of understanding with regard to the fake news phenomenon and, more broadly, in the reflexivity with regard to information behaviors. Overall, it was the older teenagers who showed the most reflexivity in their responses, thus supporting the idea of the importance of the age variable in the area of information evaluation (Metzger *et al.* 2015; Sahut and Mothe 2019).

To go further, it appears to us that two types of research could extend our work. First, quantitative research, which would allow us to consider the role of the socio-cultural environment. Indeed, quantitative surveys on juvenile use of digital technology highlight the intragenerational fault lines brought about by social positions (Hargittai and Hinnant 2008; Merckle and Octobre 2012; Cottier *et al.* 2016). On the topic of disinformation specifically, a questionnaire survey could be used to explore the relationship between young people's socio-cultural background and their perceptions of disinformation, including the amount of control they feel they have over the information environment in which it sits.

The theme of fake news, and more broadly, that of disinformation also lends itself to qualitative approaches. As is evidenced by our study, as in many others, young people frequently obtain information via their smartphones. They are therefore confronted with a particular form of "on-screen writing" (Souchier 1996) which involves a specific relationship to the information. Until now, studies on the evaluation of credibility by young people have mainly focused on information displayed on computer screens. It would therefore seem relevant to study the relationship between the specificities of reading on a smartphone and the degree of attention paid to the credibility of the information, the ways in which fake news is identified and the indicators on which epistemic judgments are based on this type of screen.

2.7. References

Aillerie, K. and McNicol, S. (2016). Are social networking sites information sources? Informational purposes of high-school students in using SNS. *Journal of Librarianship and Information Science*, 1(12), 2–12.

Allcott, H. and Gentzkow, M. (2017). Social media and fake news in the 2016 election. *Journal of Economic Perspectives*, 31(2), 211–236.

Badouard, R. (2017). *Le désenchantement de l'Internet*. FYP Editions, Limoges.

Bakir, V. and McStay, A. (2018). Fake news and the economy of emotions: Problems, causes, solutions. *Digital Journalism*, 6(2), 154–175.

Blanchet, A. and Gotman, A. (2015). *L'enquête et ses méthodes : l'entretien*. Armand Colin, Paris.

Bradshaw, S. and Howard, P. (2018). Nouvelles bidon et propagande informatique à travers le monde. In *Les fausses nouvelles, nouveaux visages, nouveaux défis*, Sauvageau, F., Simon, T., Trudel, P. (eds). Presses de l'université de Laval, Laval.

Cardon, D. (2019). *Culture numérique*. Presses de la fondation nationale de Sciences Po, Paris.

Cordier, A. (2011). Imaginaires, représentations, pratiques formelles et non formelles de la recherche d'information sur Internet : le cas d'élèves de 6ème et de professeurs documentalistes. PhD Thesis, Université Charles de Gaulle-Lille III, Lille.

Cordier, A. (2015). *Grandir connectés : les adolescents et la recherche d'information*. C&F, Caen.

Cottier, P., Michaut, C., Lebreton, S. (2016). Usages numériques et figures des lycéens au travail. In *Le lycée en régime numérique : usages et compositions des acteurs*, Cottier, P. and Burban, F. (eds). Octarès, Toulouse.

Davison, W.P. (1983). The third-person effect in communication. *Public Opinion Quarterly*, 47(1), 1–15.

Del Vicario, M., Bessi, A., Zollo, F., Petroni, F., Scala, A., Caldarelli, G., Stanley, H.E., Quattrociocchi, W. (2016). The spreading of misinformation online. *Proceedings of the National Academy of Sciences*, 113(3), 554–559.

Di Fonzo, N. and Bordia, P. (2016). Rumeurs, ragots et légendes urbaines : contextes, fonctions et contenus. *Diogène*, 213(1), 23–45.

Fallis, D. (2015). What is disinformation? *Library Trends*, 63(3), 401–426.

Flanagin, A.J. and Metzger, M. (2010). *Kids and Credibility: An Empirical Examination of Youth, Digital Media Use, and Information Credibility*. The MIT Press, Cambridge, MA.

Flechter, R., Cornia, A., Graves, L., Kleis Nielsen, R. (2018). Measuring the reach of "fake news" and online disinformation in Europe. Report, Reuters Institute, University of Oxford, Oxford [Online]. Available at: https://www.press.is/static/files/frettamyndir/reuterfake.pdf.

Gross, M. and Latham, D. (2012). What's skill got to do with it? Information literacy skills and self views of ability among first-year college students. *Journal of the American Society for Information Science and Technology*, 63(3), 574–583.

Hargittai, E. and Hinnant, A. (2008). Digital inequality: Differences in young adults' use of the Internet. *Communication Research*, 35(5), 602–621.

Harper, C.A. and Baguley, T. (2019). "You are fake news": Ideological (a)symmetries in perceptions of media legitimacy. *PsyArxiv Preprints* [Online]. Available at: https://doi.org/10.31234/osf.io/ym6t5 [Accessed 3 January 2021].

Hilligoss, B. and Rieh, S.Y. (2008). Developing a unifying framework of credibility assessment: Construct, heuristics, and interaction in context. *Information Processing & Management*, 44(4), 1467–1484.

Huyghe, F.-B. (2018). *Fake news : la grande peur*. VA Éditions, Versailles.

Jang, S.M. and Kim, J.K. (2018). Third person effects of fake news: Fake news regulation and media literacy interventions. *Computers in Human Behavior*, 80, 295–302.

Jehel, S. and Saemmer, A. (2017). Pour une approche de l'éducation critique aux médias par le décryptage des logiques politiques, économiques, idéologiques et éditoriales du numérique. *TIC & Société*, 11(1), 47–83.

Kumar, S. and Shah, N. (2018). False information on web and social media: A survey. *ArXiv Preprints* [Online]. Available at: https://arxiv.org/pdf/1804.08559.pdf [Accessed 3 January 2021].

Latzko-Thot, G. (2018). Les "fausses nouvelles", éléments d'un écosystème médiatique alternatif ? In *Les fausses nouvelles, nouveaux visages, nouveaux défis*, Sauvageau, F., Simon, T., Trudel, P. (eds). Presses de l'université de Laval, Laval.

Lazer, D.M., Baum, M.A., Benkler, Y., Berinsky, A.J., Greenhill, K.M., Menczer, F., Metzger, M.J., Nyhan, B., Pennycook, G., Rothschild, D., Schudson, M., Sloman, S.A., Sunstein, C.R., Thorson, E.A., Watts, D.J., Zittrain, J.L. (2018). The science of fake news. *Science*, 359(6380), 1094–1096.

Leeder, C. (2019). How college students evaluate and share "fake news" stories. *Library & Information Science Research*, 41(3), 1–11.

Martens, B., Aguiar, L., Gomez-Herrera, E., Mueller-Langer, F. (2018). The digital transformation of news media and the rise of disinformation and fake news. *JRC Digital Economy Working Paper*, 2018(02).

Mercier, A. (2018). Fake news et post-vérité : tous une part de responsabilité ! In *Fake news et post-vérité : 20 textes pour comprendre et combattre la menace* [E-book], The Conversation [Online]. Available at: https://cdn.theconversation.com/static_files/files/160/The_Conversation_ebook_fake_news_DEF.pdf?1528388210 [Accessed 5 January 2021].

Merckle, P. and Octobre, S. (2012). La stratification sociale des pratiques numériques des adolescents. *RESET*, 1 [Online]. Available at: http://journals.openedition.org.docelec.u-bordeaux.fr/reset/129; DOI: https://doi-org.docelec.u-bordeaux.fr/10.4000/reset.129 [Accessed 30 December 2020].

Metzger, M.J., Flanagin, A.J., Markov, A., Grossman, R., Bulger, M. (2010). Believing the unbelievable: Understanding young people's information literacy beliefs and practices in the United States. *Journal of Children and Media*, 9(3), 325–348.

Metzger M.J., Flanagin, A.J., Medders, R.B. (2015). Social and heuristic approaches to credibility evaluation online. *Journal of Communication*, 60(3), 413–439.

Pennycook, G. and Rand, D.G. (2019). Lazy, not biased: Susceptibility to partisan fake news is better explained by lack of reasoning than by motivated reasoning. *Cognition*, 188, 39–50.

Proulx, S. (2018). L'accusation de fake news : médias sociaux et effets politiques. In *Les fausses nouvelles, nouveaux visages, nouveaux défis*, Sauvageau, F., Simon, T., Trudel, P. (eds). Presses de l'université de Laval, Laval.

Rebillard, F. (2017). La rumeur du Pizzagate durant la présidentielle de 2016 aux États-Unis. *Réseaux*, (202–203), 273–310.

Revault D'Allonnes, M. (2018). *La faiblesse du vrai : ce que la post-vérité fait à notre monde commun*. Le Seuil, Paris.

Sahut, G. (2014). Les jeunes, leurs enseignants et Wikipédia : représentations en tension autour d'un objet documentaire singulier. *Documentaliste – Sciences de l'Information*, 51(2), 70–79.

Sahut, G. (2017). L'enseignement de l'évaluation critique de l'information numérique : vers une prise en compte des pratiques informationnelles juvéniles ? *TIC & Société*, 11(1), 223–248.

Sahut, G. and Mothe, J. (2019). Epistemic vs non-epistemic criteria to assess Wikipedia articles: Evolution of young people perception. *Information Literacy in Everyday Life: 6th European Conference on Information Literacy 2018, Oulu, Finland, 24–27 September 2018*, Revised Selected Papers, Berlin. Springer, 329–339.

Serres, A. (2007). Questions autour de la culture informationnelle. *The Canadian Journal of Information and Library Science*, 31(1), 69–85.

Serres A. (2012). *Dans le labyrinthe : evaluer l'information sur internet*. C&F, Caen.

Søe, S.O. (2016). The urge to detect, the need to clarify. Gricean perspectives on information, misinformation, and disinformation. Unpublished doctoral dissertation, University of Copenhagen [Online]. Available at: https://static-curis.ku.dk/portal/files/160969791/Ph.d._2016_Obelitz.pdf [Accessed 3 January 2021].

Souchier, E. (1996). L'écrit d'écran, pratiques d'écriture & informatique. *Communication & Langages*, 107(1), 105–119.

Starbird, K. (2017). Examining the alternative media ecosystem through the production of alternative narratives of mass shooting events on Twitter. Communication, 11th International AAAI Conference on Web and Social Media, Venice, 230–239.

Starbird, K. (2019). Disinformation's spread: Bots, trolls and all of us. *Nature*, 571(7766), 449.

Tandoc, E.C., Lim, Z.W., Ling, R. (2018). Defining "fake news". *Digital Journalism*, 6(2), 137–153.

Tricot, A., Sahut, G., Lemarie, J. (2016). *Le document : communication et mémoire*. De Boeck Supérieur, Louvain-la-Neuve.

Vermersch, P. (1994). *L'entretien d'explicitation en formation continue et initiale*. ESF, Paris.

Vosoughi, S., Roy, D., Aral, S. (2018). The spread of true and false news online. *Science*, 359(6380), 1146–1151.

Waszak, P.M., Kasprzycka-Waszak, W., Kubanek, A. (2018). The spread of medical fake news in social media – The pilot quantitative study. *Health Policy and Technology*, 7(2), 115–118.

Zannettou, S., Caulfield, T., De Cristofaro, E., Kourtellis, N., Leontiadis, I., Sirivianos, M., Stringhini, G., Blackburn, J. (2017). The web centipede: Understanding how web communities influence each other through the lens of mainstream and alternative news sources. In *Proceedings of the 2017 ACM Internet Measurement Conference*, Uhlig, S. and Maennel, O. (eds). Association for Computing Machinery, New York.

Zannettou, S., Sirivianos, M., Blackburn, J., Kourtellis, N. (2019). The web of false information: Rumors, fake news, hoaxes, clickbait, and various other shenanigans. *Journal of Data and Information Quality*, 11(3).

3

"A Big Nebula that is a Bit Scary" (Louise, Trainee Schoolteacher): Training through/in Digital Technology, in School and in Professional Training

In 1986, Ulrich Beck crowned the advent of a so-called "risk society", which is no longer solely concerned with the endangerment of individuals by forces of nature, but sees technological development as a source of deep insecurity for individuals and the collective in general (Beck 2001). Today, many digital risks have been identified, covering fields as varied as computer science, law, cognition and information. The media panics linked to these digital risks are knocking on the doors of schools[1], which are being called upon to adapt to the digital world. This in turn calls into question the teachers, who are themselves social actors "caught", in the Sartrean sense of the term, in the demands, fears and hopes stirred up by society (Sartre 1985), as well as their institution of belonging (Cordier 2018), and the very exercise of a profession that consists of "acting in urgency (and) deciding in uncertainty" (Perrenoud 1999). The teaching profession is fundamentally a profession in which risk-taking reigns, with pedagogy calling for a consideration of both otherness and alteration, welcoming knowledge that is always in motion, made up of permanent re-configurations (Veyrié 2014).

Chapter written by Anne CORDIER.

1 A prime example of this is the anxieties generated, in particular among parents, by a TV documentary broadcast on January 18, 2018 as part of the Envoyé Spécial program, which associated autism with screen time based the claims made by Dr. Anne-Lise Ducanda.

This risk-taking seems to be strongly accentuated when it comes to these actors taking on a mission that has now been fully devolved to them, namely education through and in digital technology: the reference framework of competencies for teaching and education professions, dated July 2013 in relation to the law on course guidance and study programs to rebuild the school system in France (*Loi d'orientation et de programmation pour la refondation de l'École de la République*), stipulates that teachers must "integrate the elements of digital culture necessary for the exercise of the profession", which implies both "making the most of the tools", and helping students to use them in a digital context.

While teaching is already a risk in itself, as evidenced by theoretical work on the act and circumstances of teaching, as well as by the testimonies of teachers on this subject, what is the situation – at a time when pervasive media panics cohabit with techno-determinist myths – of training through and in digital technology, whether in the context of the classroom or in that of professional training[2]?

Anxious to grasp and understand the practices and representations of digital technology, and more broadly, of media and information literacy (MIL) among trainee teachers and their trainers, the researchers and practitioners involved in the Prémices project[3] deployed a mixed methodology, combining quantitative and qualitative data collection within the same ESPÉ (French teacher training institute)[4]. In concrete terms, a questionnaire distributed between February and April 2017 made it possible to question – on a voluntary basis – 243 trainee teachers (ST, HMT, PHT,

2 In order to facilitate understanding of the teaching and learning spaces referred to in this contribution, we will refer to "classroom practices" to designate the practices of teachers working in the first and second levels of education, and to "practices in professional training" to designate the practices of teachers working in the ESPÉ.

3 This chapter, written by Anne Cordier, is the result of data collection and analysis work carried out collectively by the Prémices Project Group, which is composed of: Brière Hélène, Delhaye Cyrille, Dequin Thomas, Ermel Laurence, Flutre Isabelle, Gouello Karina, Manoury Perrine, Metterie Sandrine and Pelletier Béatrice. All members of the Project Group participated in the review of this chapter.

4 *École Supérieure du Professorat et de l'Éducation*. It should be noted that we have chosen to keep the acronym ESPÉ at a time when the INSPÉ (*Institutions Nationaux Supérieurs du Professorat et de l'Éducation*) have succeeded the ESPÉ. This choice is motivated by the fact that the data collected was done so within the geographical and political framework of an ESPÉ.

PEC[5]) and 54 trainers (all statuses combined) from this institution, on their personal practices of information and digital technology, as well as their pedagogical practices, and the professional training experienced or implemented. Following the quantitative survey, semi-structured interviews were conducted in pairs, in May–June 2017, with 10 volunteer trainee teachers. This research-for-action project, focused on information cultures which, in our view, partially encompass digital culture (Baltz 1998, 2015; Cordier and Liquète 2014), is original in its adoption of an ecosystemic approach, considering, in an interactional dynamic, the action of initial (professional) training and the action of training in the classroom: training through digital technology and digital technology is, or should be, a challenge for trainers in the ESPÉ to develop the professional skills of young teachers who will have to experience digital technology as a tool and as an object of teaching and learning in the classroom.

3.1. Social beings, above all else

While many societal and pedagogical discourses, supported by scientific research, constantly remind us that "the student is a person" (Perrenoud 1999) and that the social and cultural context, as well as the context of their extra-academic activities, must be taken into consideration (Dioni 2008; Cordier 2011), it is worth noting that the teacher is still too often considered as an academic being with a circumscribed institutional role. However, understanding the personal biography of the teacher seems to us to be crucial in order to grasp how they approach the act of teaching and learning, both emotionally and pragmatically. For this reason, our research project has focused on drawing a sociological and informational portrait of the respondents, both within and beyond the academic context.

3.1.1. *A "fluid identity" to be grasped*

The notion of "fluid identity" has thus deeply resonated with our theoretical and methodological reflections, and subsequently with the data collected as a result of our investigations. It is François de Singly who, as early as 2003, invites us to consider the actor as the bearer of a personal and social history, and of a repertoire of roles, which they engage in within

5 ST = schoolteacher; HMT = high school and middle schoolteacher; PHT = professional high schoolteacher; PEC = principal education counsellor.

different social contexts. From this perspective, it is not a question of thinking about the actor's behavior within differentiated social universes – and of considering it as relative to these universes – but of conceiving and grasping the multidimensionality of individual identity within a resolutely continuous process. This multidimensionality is observable, for example, through places: de Singly takes the example of a teenager's bedroom, which contains objects referring to several dimensions of their identity: the space of the son or daughter, the space in which the student's schoolwork is carried out, as well as the space which is home to youth culture (de Singly 2003). The same is true for the teacher, who at the heart of their professional practice, in the classroom, is also a social being, inhabited by a personal history, taking on the role of parent if necessary. We were very concretely confronted with the operationality of this theory of fluid identity, by identifying the "double I" (de Singly 2017) of the respondents. Thus, it is by referring to his personal digital practices and his relationship to digital technology as a parent for his own children, that Loïc, a teacher in Engineering Sciences, answers our questions on digital technology at school:

> There's nothing to worry about, I'm sure you have to use all of the tools you have and everything as much as possible, it's fine. But for small classes, I'm not sure [...] Because I have children too, and I wonder when to start introducing them to these (Loïc, THST (technical high schoolteacher), Computer Science, 38 years old).

In the same way, Eulalie, a schoolteacher, reflects on her pedagogical relationship with digital technology in light of the practices observed around her younger brother:

> I see my brother who is younger… (Eulalie, fourth-grade ST, 23 years old).

It goes without saying that the trainers working in the ESPÉ also bring their many social roles to the training space.

The research protocol set up within the framework of this Prémices project therefore retained from the outset, the principle of investigating the fluid identity of the respondents. Sociological data (age, family and cultural environment) complemented the data from the so-called non-formal sphere and the so-called formal, academic sphere.

In concrete terms, the main lines of questioning concerned the personal information practices of the actors (trainee teachers and trainers), their perception of the contemporary issues of media and information literacy and their positioning as education professionals with regard to informational, media and digital objects in training (trainee teachers with regard to their students, trainers with regard to trainee teachers), as well as their view of MIL and digital training in ESPÉ.

The following is a graphical representation of the age profiles of the actors surveyed through the questionnaire. It should be noted that these are not the complete age profiles of the trainee teachers (FSTG) and trainers in the field, but the profiles of those who agreed to participate in the survey.

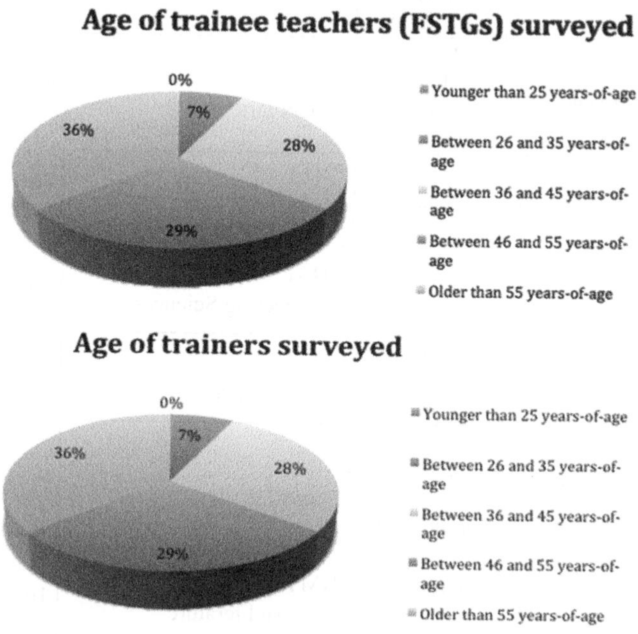

Figure 3.1. *Age of the respondents. For a color version of this figure, see www.iste.co.uk/capelle/digitalrisks.zip*

Table 3.1 summarizes the profiles of the student teachers who participated in the qualitative survey.

First name	Age	Social status	Body Discipline	Teaching level
Annaick	27 years old	In a relationship and living with a partner	ST Kindergarten	Pre-K[6]
Eulalie	23 years old	In a relationship and living with a partner	ST Second grade	Fourth grade
Hélène	36 years old	In a relationship, civil partnership	PHT Ninth grade	First STMG
Laeticia	44 years old	Married, three children (teenagers)	ST Kindergarten	Pre-K
Léa	28 years old	Single	PHT Biotechnology: Prevention Health Environment	Second Year Pro Loïc
Loïc	38 years old	Married, two children (primary)	THST Engineering Sciences	First S, Tale S
Louise	39 years old	Married, one child (primary)	ST Second grade	Second grade
Mathilde	24 years old	In a relationship, civil partnership	ST Second grade	Fourth grade
Nina	24 years old	In a relationship, civil partnership	HMT Modern literature	10th grade
Vanessa	32 years old	Single	HMT Information and resources	High school

Table 3.1. *Profiles of the student teachers surveyed*

6 Pre-K = pre-kindergarten, ST = schoolteacher, HMT = high and middle schoolteacher, PHT = professional high schoolteacher, SHT = senior high schoolteacher.

3.1.2. *Digital technology in the actors' personal ecosystem*

Taking an interest in the role of digital technology in the personal ecosystem of the actors surveyed is essential in order to be able to address, in the most complex way possible, the understanding of digital technology in the classroom and in professional training by these same actors. Sallaberry's work has shown how much the representations of the object to be taught, the students' practices and the teaching and learning act influence the practices, views and gestures adopted by the teacher in context (Sallaberry 1996). Representations make it possible, in all situations, to understand our personal relationship to the world and to situate ourselves within a social group. They thus form the basis of a mental construction, "more or less affectively charged, made from what the person has been and what they project, they guide their action and lead to the behavior they will adopt" (Postic and De Ketele 1988, p. 37). The teacher thus grasps the world around them according to their beliefs, values and opinions.

Understanding the representations and non-formal practices of teachers and trainers in terms of the use and understanding of digital tools within their personal sphere is imperative to gain a better understanding of their pedagogical positions and choices. On the strength of previous research showing the influence of digital representations (Dioni 2008), as well as the actors' sense of personal expertise on the understanding of the digital object in a teaching and learning situation (Cordier 2010, 2012), we questioned the respondents on their sense of expertise concerning "the digital realm".

On a scale of 1 (novice) to 5 (expert), we noted that 80% of both trainee teachers and trainers are between level 3 and level 5, sharing a relatively high level of expertise in relation to digital technology. However, we can hypothesize that, particularly concerning the trainers in the ESPÉ, those who responded to the quantitative survey have a more peaceful relationship with digital technology than those who did not voluntarily participate.

However, the responses between the types of actors markedly differ when we go into more detail about their personal use of digital technology, and in particular, the possession of an account on a digital social network (DSN). While 91% of the trainee civil servants in our study said they had an account on a social network, 42% of the ESPÉ trainers said they had one.

It seems to us that this difference in the number of accounts on DSNs between trainee teachers and ESPÉ trainers foreshadows the non-use of DSN, particularly non-professional ones, and their non-use for pedagogical purposes in the context of training and then in the classroom. Furthermore, we note that Facebook is very much at the top of the list of DSNs employed for personal purposes by teacher trainees and trainers: the virtual absence, in the results, of accounts held on professional DSN prescribed by the institution (*Viaeduc*, for example), whether among teacher trainees or their trainers, calls into question the process of professional socialization (Dubar 1991, 2005) intentionally underpinned by these tools, as well as, of course, the support for this process by the institution itself.

3.2. Understanding of digital technology in the classroom

It is with a singular history, a sense of personally assessed expertise and equally personal practices deployed in the non-formal sphere that teachers approach the digital world in the classroom and in the training context. We thus sought to understand how trainee teachers envision the use of digital technology in their classroom practices with students, both politically (underlying ideology), emotionally (Déchaux 2015) and pragmatically (implementation)[7].

3.2.1. *Crystallization and awareness of issues*

There is no doubt that trainee teachers are aware of the issues related to digital education for students in primary and secondary schools. 86% of them even declare that they feel "personally concerned by taking charge of the digital training of students"[8].

All of the interviewees expressed an awareness of an educational responsibility in the face of digital risks, which appear at the heart of their concerns when anything "digital" is mentioned. First of all, it is a question of

7 It should be noted that in the absence of detailed qualitative data collected on the experiences and pedagogical practices of the ESPÉ trainers, we prefer to focus on the trainee teachers in this section. It is important to not draw hasty conclusions from quantitative results alone, which are fragmentary not only because of their declarative nature, but also because of the limited number of ESPÉ actors who responded to the survey.

8 The wording used by the Prémices group in the questionnaire submitted to respondents.

responding to an institutional demand, very present in their discourse through the third party-absent analyzed by Charaudeau (2004): "*We're told all the time, we're told that we have to teach students that there are silly things on the Internet, but we also have to be careful not to say that everything on the Internet is fake*", insists Mathilde (fourth-grade ST, 24 years old). Louise (second-grade ST, 39 years old) says it just as clearly, evoking the requirement to use digital technology as a teaching tool: "*You have to do digital*".

These trainee teachers, who are often younger than the average teacher in their institution, consider that they have an even greater role to play because they belong to a "*digital generation*", as Léa (Biotechnology PHT, 28 years old) puts it. Several of them confide in us the division between "*old teachers*" and "*young teachers*", like them (Eulalie, fourth-grade ST, 23 years old). To illustrate this point, Mathilde takes the example of the blockages in her school around the digital school report book: "*I'm laughing because it will be difficult with my colleagues! [...] They have their little quiet life*" (Mathilde, fourth-grade ST, 24 years old). Mathilde verbalizes one of the reasons why it seems incontestable to us to consider, from the point of view of the teachers, the integration of digital technology as a risk: to use digital technology in our professional practices is, to a certain extent and for certain actors, to leave a comfort zone established over time, in particular through the establishment of routines.

The young teachers in our study are perfectly aware of this and consider that, given their age and this moment of entry into the profession, they cannot avoid this institutional demand. For the demand is not the only reason for their motivation: they are aware that they have an educational responsibility. This responsibility can be broken down into four major arguments. The first argument is that of dealing with a public that needs digital technology for its personal and professional fulfillment:

> We can hardly do without (digital), especially nowadays, given that our students are from the digital generation (Eulalie, fourth-grade ST, 23 years old).
>
> Because they (the students) are going to work in a digital environment (Laeticia, kindergarten ST, 44 years old).

The second argument is that of dealing with a public that needs specific digital skills: the skills of searching for information, evaluating sources and information, and understanding information are thus cited. The third argument pushes these teacher trainees to feel responsible for an education through/in digitalization:

> We can't shield our classroom because they have all of this outside anyway [...] The school has to be somewhere they can discuss all of this, because they don't necessarily do so with their parents (Eulalie, fourth-grade ST, 23 years old).

In their view, it is their responsibility to discuss their digital practices with their students in order to make them aware of them, so that their students "*don't get manipulated too much*" (Loïc, THST, computer science, 38 years old). Finally, the fourth argument used by these young teachers is the "risks" to which the youngest are exposed, particularly in their digital use, with the problems of "cyber-bullying" and "screen addiction" being given priority when it comes to dealing with digital risks from an educational perspective:

> They never, they have nothing to tell us in the morning. They don't live much behind their screen. Mathilde (second-grade ST, 24 years old) laments.

3.2.2. *When the socio-technical framework hinders the entry of digital technology into the classroom*

All of the trainee teachers, and particularly those working in primary education, denounce dilapidated, inadequate or completely absent computer equipment. It is obvious that this inadequate socio-technical framework hinders the pedagogical practice that is so desired by the institution and by these young professionals. In view of the computer equipment in her school, Eulalie says: "*Digital technology is not a current issue for (her) school!*" (Eulalie, fourth-grade ST, 23 years old). This technical under-equipment is even a recurrent subject of complaints between trainee teachers.

From Mathilde, who was never able to use the video projector in the computer room "*because the remote control was missing*", to Laeticia, who found herself confronted with an interactive whiteboard on which her colleagues had simply "*stuck a white marker pen on it to write on*", to Annaïck, who, in order to carry out the "*digital session*" required as part of her initial training, had to scroll through each pupil on the only computer in the class (26 children in kindergarten), in order to enter their first names; it is clear that trainee teachers often work in highly unfavorable technological contexts.

If access to a reliable technical framework and sufficient equipment is not always present in secondary schools, some respondents also point out the demanding and useless character of massive equipment. Like Léa, who ironically denounces the arrival of tablets in her school:

> The arrival of the tablets, it made me laugh a lot. In the whole school, all of the kids had to have a digital tablet [...]. It's just spending money, in my opinion! And I think that in, come on, in five years, the tablet will be a has been! And when I say has been, it'll be like the Minitel is today! (Léa, 10th-grade PHT, 28 years old).

We can see here, the gap between an institutional demand to use digital technology and equipment that is either deficient or considered self-fulfilling by the institution: it is not enough to provide equipment in order to modify pedagogical practices, and the examples cited by the trainee teachers, ranging from sarcasm to disappointment, are very eloquent in this respect.

In fact, the fragility of the equipment available in the schools is an obvious obstacle to a peaceful understanding of digital technology in teaching practices. The risk-taking is all the more important for these young teachers, who are aware that they will have to face technical difficulties, manage the class and master the didactic scenario developed beforehand:

> It's true that it doesn't sit right to say, well, I'm going to have to go digital, and that's going to be problematic again [...]. It's complicated to manage, and the problems of digital that the students encounter, because it disconnects, because they can't find it, because... (sighs) and the pedagogy (Laeticia, kindergarten ST, 44 years old).

3.2.3. *Rather modest and low-risk experiments*

Despite this uncomfortable socio-technical framework, what about pedagogical practices engaging digital technology in the classroom?

First of all, it is difficult to effectively measure the use of digital technology by trainee teachers in their teaching practices. In fact, 100% of them have implemented "a session based on the use of digital technology" during the year of training at the ESPÉ, this session being imposed in the training curriculum and the evaluation of the trainee teachers (FSTG). The institutional prescription thus masks the realities of voluntary support for trainee teachers. Moreover, "digital use" covers a very broad spectrum, ranging from the use of a slide show to the facilitation of a reading session on tablets, and including training in digital identity. The designation of "digital pedagogical practices" is so uncertain, and the punctuality of this exploitation so strong, that we note that none of the 12 interviewees considered that they had used digital technology to teach in this year of the course!

More specifically, we have certainly encountered young teachers who are reluctant to use digital technology in the classroom for ideological reasons. Loïc (THST, computer science, 38 years old), for example, makes a point of defending "pedagogy without digital technology" (but we note in the account of his experiences in the classroom that he exchanges with his students on the subject, notably concerning "health risks"). Mathilde (fourth-grade ST, 24 years old), who has already experimented with digital technology, admits that she is reluctant to embark on digital technology in the classroom. Both are afraid of participating in the accentuation of intensive, even addictive, screen practices of their students:

> I always think about it when I start thinking about a computer session, I'm always thinking "but you're going to do an hour of computer, is that really smart, knowing they're going to spend their evening on video games?" Honestly! (Mathilde, fourth-grade ST, 24 years old).

But the majority of young teachers with whom we spoke expressed the need and even the "desire" (the term was used spontaneously by the respondents concerned) to integrate digital technology into their classroom practices.

However, given both the unfavorable socio-technical frameworks within schools, and the fear represented by the uncertainty of using technology in the classroom, pedagogical experiments involving digital technology appear to be "low-risk", to use Vanessa's term (HMT, information and resources), allowing young professionals to remain in a certain comfort zone. It is an education through digital technology that is quantitatively more important. It is above all a question of supporting disciplinary learning through education using documents (not formulated as such by the respondents, and therefore not necessarily conscious or verbalized to the learners in the situation). Eulalie, who worked in a fourth-grade class this year, is a young and extremely enthusiastic teacher who also has very intensive professional information monitoring practices. The findings from this personal monitoring feed her practices and pedagogical scenarios. She uses a lot of video documents to develop dialogue with her students and the understanding of disciplinary content. Léa adopts the same logic with her students in vocational high school, using documents to "illustrate the course" (Léa, Biotechnology PHT, 28 years old). Nina, a literature teacher, had her students in 10th grade do research on a literary movement or authors, but she took care to indicate to them "beforehand the sites they should visit because (she did not want them) to copy from Wikipedia", she specified (Nina, Literature HMT, 24 years old). The other respondents favored manipulative learning, such as keyboard entry (Annaïck and Laeticia, kindergarten ST, 27 and 44 years old, respectively; Louise, second-grade ST, 39 years old), the use of spreadsheets (Hélène, PHT Tertiary, 36 years old) or the processing of computer files:

> I have a student who has a YouTube channel. It amuses me a lot, because when I tested their word processing skills, there's nothing, they can't create a document, they can't do save as. On the other hand they have a YouTube channel! (Mathilde, fourth-grade ST, 24 years old).

As for digital education, although it is an integral part of the institutional requirements within the school programs, it seems to be relegated to extremely punctual sessions during the school year, or even one-offs. In all cases, it is the "*dangers*" linked to digital technology that are at the forefront of the experiments conducted. The legal risks on the web, and particularly the question of digital identity, are a major concern for these trainee teachers. Eulalie carried out training for her class of fourth graders, based on the online video resource "Vinz et Lou", produced by *Internet sans Crainte*.

Hélène, PHT Tertiary participating in Economics, Law and Management with a class of first STMG (Penultimate year of the Sciences and Technologies of Management section of the technological baccalaureate), set up a session of PS (personalized support) based on a serious game – whose name she does not remember – in order to approach "*the precautions to take with regard to one's identity and one's digital footprint*". These are the only examples of pedagogical experimentations with digital technology as an object of study, mentioned by the civil servant trainees interviewed. However, it is worth recalling the potential gap between what is verbalized because it is conscious and what is actually carried out; it is possible that by observing these young teachers in the classroom, we have, for our part, identified time devoted to digital technology as an object of study. The protocol set up within the framework of this project does not allow us to state this, being limited to declarative practices.

3.3. Teaching with and through digital technology: Constant risks

The PRÉMICES project aims to forge connections between pedagogical practices that use digital technology as a tool and as a teaching and learning object in the classroom, and practices that have been tested in professional training. Teaching with and in digital technology is probably a point of tension, both in the classroom and in pre-service training; gaps are emerging between the views and needs expressed by young teachers in training and the positions of ESPÉ trainers. Our aim is to examine the resources – both obstacles and levers – that are conducive to the transition from information to knowledge, while at the same time making the hypothesis that without action, knowledge is not very efficient in pedagogy and training.

3.3.1. *Tensions in the classroom*

As we have seen, within the classroom, the trainee teachers first of all point to a training ecosystem that is complex to grasp and master when digital technology is included. But if the teaching and learning situation is made more complex by the use of digital technology because of the fragility of the socio-technical framework itself, it is also made more complex, according to these young teachers, because digital technology raises profound questions about teaching authority and adds difficulty to an already crucial search for the right stance on this in the classroom. The testimonies of Eulalie and Louise, both schoolteachers, summarize the fears expressed by the respondents:

> You have to be able to respond… so it has to be even more in the second grade, I guess in kindergarten well, you can still get past them without a problem (Eulalie, fourth-grade ST, 23 years old).

> It's hard to explain, but in fact, teaching a lesson in French, putting together clips in French, in math, it's… it doesn't scare me… Putting together clips in digital technology, how do you take it, you say to yourself: "I have to re-orient my objectives and my skills. How am I going to direct it? How am I going to sell it? How am I going to present it to them?", because when you don't master it, they sense it very quickly. So, I needed time, extra time in my head to draw things and build them, whereas, in more traditional disciplines, I didn't need that time (Louise, second-grade ST, 39 years old).

"Having the ability to respond", to use Eulalie's expression, means knowing how to deal with content that has not yet been mastered (or that is felt to have been mastered), and also with a teacher/student relationship that is difficult.

Vanessa, Information and Resources HMT, admits that she herself felt helpless when faced with certain technical questions from students, asking how the Internet, fiber optics, broadband, etc., work. As a student in the Master MEEF Resource Management program, with general training general Information and Communication Sciences and specialist training in Information and Resource Management in particular, she recognizes:

> I gave a kind of summary response, but I realized that I had to have a look at myself, really, that I had to take the time theoretically to… to almost teach myself a course like we had in the first year of our master's course, so something which is very in-depth, so that I could be prepared for these questions (Vanessa, Information and Resources HMT, 32 years old).

This demand for expertise in the field is all the more frightening for the other respondents because they have no disciplinary training in digital, informational and media studies.

In addition, many of them point to the porosity caused by the digital world between the domestic, intimate sphere of the student and the school sphere; a porosity that requires them to position themselves as educators, which is a role that they do not feel capable of assuming. On the one hand, these teachers realize that their personal digital practices as social actors are in some way accessible to students. Thus, Mathilde returns to an episode experienced outside of the classroom which questions the way in which she constructs her own digital presence on DSN in the light of her personal practices and her professional life:

> I have a student who came to me and he said to me, plain as day: "I went on Facebook with my mom and we tried to search for your Facebook profile. We didn't find it but we found the one of your group" (Mathilde, fourth-grade ST, 24 years old).

This experience had a strong impact on the young teacher, who felt, in her own words, "vulnerable" in the face of this student sharing with her this search for a digital identity. On the other hand, they also see the students' personal digital practices – about which they often express harsh judgment – coming into the classroom, along with their consequences. For example, Eulalie recounts that she has often had to manage conflicts between students that arose during non-formal activities on social networks, or simply had to understand events that took place outside of school time on these sites; in her view, this "*complicates relations within the class*" (Eulalie, fourth-grade ST, 23 years old).

Finally, it is the register of fear that dominates the discourse of these young teachers, when it comes to evoking their understanding of digital technology in the classroom.

From the outset, Louise tells us: "As a trainee teacher starting out, saying you're going to implement 'digital technology with students' is a big statement. It's the slightly scary side to it, really, that you have to cover digitalization" (Louise, second-grade ST, 39 years old). However, she later recognized how much she realized the fantasy she had finally nurtured regarding digital resources in the classroom, imagining setting up extremely complex projects:

> At the beginning it's a bit of a big nebula and it's a bit scary. In the end, by taking on smaller projects, by setting up small projects, it remains a means and something very pleasant to work on, since it allows us to approach them in a different way (Louise, second-grade ST, 39 years old).

Despite this remark, which testifies to a change in both the pragmatic and emotional understanding of the object in her teaching practice and teacher–student interactions, Louise's general discourse throughout the interview is characterized by the lexical field of anxiety. The fear of not being up to par pedagogically haunts all of these trainee teachers. A fear of which they are perfectly aware, and from which they seek to free themselves. Eulalie admits that:

> There are so many risks and that scares me already, and I think that there is also this culture of fear that comes with the digital realm since we start off by saying "be careful!", and that's what can scare some people (Eulalie, fourth-grade ST, 23 years old).

Her own fears guide her understanding of digital technology in the classroom, but she does not succumb to the hasty conclusion to renounce digital uses in the classroom because of the potential risks. The young teacher admits that she is careful not to "over-constrain (her) students" because she herself is afraid of the consequences of browsing, or of using tools such as DSN that she does not frequent or master in her opinion.

3.3.2. *Tensions in training*

How can we explain the fact that these young teachers are thus imbued with fears about an object that they are nonetheless mostly using, sometimes intensively, in their daily lives? Research has already shown that there is a mismatch between engagement in daily digital practices and engagement in digital uses in education (Rinaudo and Ohana 2009; Cordier 2011; Capelle *et al.* 2018). The PRÉMICES project looked at the point of view of ESPÉ trainers, in order to seek complementary keys to understanding.

If it is problematic for trainee teachers in the classroom, the socio-technical framework is also a source of tension within the value system of the actors in professional training. Thus, questioning trainee teachers and ESPÉ trainers about the use of their students' personal smartphones and/or tablets in class brings to light a significant disparity: while 91% of the trainee teachers surveyed said they were in favor of the use of students' smartphones and/or tablets in class, as part of a pedagogical project, only 44% of the ESPÉ trainers allow/would allow their students in training to use these connected objects.

This same disagreement appears, although less marked, concerning the use of social networks with learners in class: more than 80% of the trainers who responded to the questionnaire are opposed to it with their students, while nearly half of the trainee teachers are in favor of this use with their students in class.

These disagreements raise the question of how to encourage the transfer of practices from training to the classroom when the practices have not been tested in the context of training, which is reassuring.

This question is all the more acute if we judge how trainee teachers view the training they receive during their year of responsibility.

Thus, 42% of the trainee civil servants surveyed consider that the so-called "Digital Culture Core Curriculum" (DCCC) courses given at the ESPÉ are adapted to their needs for training with digital technology. Only 29% consider that these courses meet their needs for training with digital technology. It should be noted that among the respondents, 32% responded that they were not concerned by these questions, as they belonged to courses that were part of the DCCC.

There are several explanations for these quantitative results. First of all, it is important to specify that the "Digital Culture Core Curriculum" (DCCC) courses are given to all of the master's courses in teaching, education and training (*master métiers de l'enseignement, de l'éducation et de la formation*), for both primary and secondary school levels. However, several of these courses have not included this DCCC in their curriculum because they have taken on the digital issue within the course itself. However, students, during their interviews and informally, revealed that the hours officially devoted "to digital technology" in these courses offering the

DCCC were in fact replaced by hours dedicated to "pure disciplinary didactics" (as one interviewee put it), that is, the course's core specialist subject, without including digital technology in any way whatsoever. Some teacher trainees are therefore excluded from training within digital technology as part of their initial training, as these hours are repurposed by the trainers themselves[9]. This repurposing of institutional requirements, in particular, seems to us to be a strong indication of a "disciplinary conscience" (Reuter 2007) that is undermined in the eyes of ESPÉ trainers by digital resources, and a desire to preserve the discipline and the related hours, in the absence of a vision of education which incorporates training through/in digital resources. Secondly, if 42% of the trainee teachers surveyed find that the DCCC courses are suitable for training with digital technology, it is because an approach centered around tools and operations is favored. The interviews conducted with the trainee teachers (FSTGs) on this subject give rise to a catalog of tools presented without any pedagogical contextualization, each one stating that it is ultimately up to them to use them… or not. It is easy to imagine that a professional training student who is already highly reluctant to use digital technology in their classroom practice will not, of their own accord, engage in self-taught practice by devoting additional time to understanding the pedagogical and didactic issues of the tool presented in training. Finally, the degree of satisfaction expressed by trainee teachers with regard to digital training is even lower, with less than a third of respondents believing they are able to "train in digital technology". As we saw earlier, the requirement to master content is one of the main reasons why trainee teachers do not engage in, or fear engaging in using digital resources with their students.

3.3.3. *Desires on both sides*

The picture painted may seem bleak. However, we have met a number of trainee teachers and trainers who are genuinely interested in embracing digital technology as both a tool and an object of teaching and learning.

When we interviewed the young teachers at the end of the practicum year, we were able to determine that there were significant frustrations

9 The point here is not to judge who makes these choices. Let us not overlook an important contextual fact: the integration of digital technology into primary and secondary teaching has extremely varied outcomes depending on its contribution (Lehmans *et al.* 2017).

regarding unrealized uses of digital technology. Laeticia would have liked the interclass exchange she set up with her kindergarten students to be based on digital communication, but this was not possible. Mathilde expressed a similar disappointment:

> At the moment, when we want to do document-based research etc., we pre-select a book for them that we borrow from the library with a view to letting them go and look on the internet independently. But the computer room is too far from my classroom, and I don't have access. I can't see the computer room well enough to supervise them so I can't send them. I'm not allowed to (Mathilde, fourth-grade ST, 24 years old).

"The desire to test other things" (Eulalie, fourth-grade ST, 23 years old) is certainly present, and those who, during this year of training, wanted to use digital technology in the classroom and experimented, even if they were unsuccessful, unanimously expressed the desire to go further and continue their pedagogical experiments. Mainly, the young teachers emphasize the prevention of digital risks for which they feel responsible with regard to their students, following the example of Mathilde (fourth-grade ST, 24 years old), who is concerned about working on the relationship between young people and screens (limiting "screen time", in particular). But these trainee teachers also want to overcome their own fears in pedagogy, the most important of which is the use of DSN for teaching. Eulalie, for example, is very interested in Twictées, and more generally, in collaborative digital writing. It is also about taking more risks in our own classroom practice. Vanessa, a junior high schoolteacher in charge of information and resources, shares the sessions she has set up this year. She speaks of the progression in information-documentary work that can be described as "traditional", made up of an understanding of the keys parts of the book, an awareness of the media and its history. And during the interview, she spontaneously admits:

> In retrospect, I regret having done that because it was "easy" [...]. I didn't challenge myself by saying to myself "they are in sixth grade", but I can totally start talking about… well, that's something else entirely, but still, about digital identity for example [...]. Well, for many of them they already had a Facebook account, even though they are not old enough to have

> one legally. So, we were able to come back to this a little bit because we discussed it. And, of course, they need these things because they also get their information from Facebook, so I could have very well put together a sequence on that. But is it because it was my first year, I don't know, but I needed to do little things that I really felt capable of at that point. And I think I needed to succeed too and not necessarily get in trouble. So, I kind of regret not being more ambitious or more courageous (Vanessa, Information and Resources HMT, 32 years old).

Aware that this "*ambition*" of which Vanessa speaks, and more generally, this risk-taking in pedagogy, must be encouraged in professional training, trainers are deploying initiatives within the ESPÉ. It is a question of integrating digital technology into the daily training of trainee teachers in a way that is tenuously linked to their specific discipline(s). The implementation of an *escape game* in the teacher training master's course in Music Education[10], as well as in the teacher training master's course in Special Educational Needs[11], or the use of interactive whiteboards or online quizzes, are likely to encourage pedagogical practices based on the relaxed and reflective use of digital technology in the classroom[12]. To date, we do not have enough information on these experiments to measure their actual impact on the representations and practices of the trainee teachers who benefit from them. It should be emphasized that these are initiatives that aim to promote the use of digital technology as a means of teaching and learning, and not as an object integrated into media and information literacy.

---

10 Video presentation of the escape game "*Franz a disparu*", created for the teacher training master's course in Music Education by Sandrine Metterie (trainer and co-leader of the course) and Adeline Mahieu (Espé digital educational engineer). See: https://www.youtube.com/watch?v=AN9Wj1VstbI.

11 Within the ESPÉ in Rouen, a working group on escape games is being set up, demonstrating the awareness of trainers to integrate playful and digital practices into professional training. In addition, in September 2019, a PédagoLab will be set up at the INSPÉ in Rouen, under the direction of Anne Cordier, with the aim of supporting the use of digital technology in training, encouraging the use of active teaching-learning methods, based, in part, on the use of digital technology, and trainers in digital culture and informational and communicational issues.

12 It should be noted that it seems impossible to have a completely exhaustive idea of what is being done with digital technology within the training courses at the ESPÉ. The examples given here are those which we are aware of, and do not reflect all the pedagogical practices in training.

3.4. Potential courses of action

Through the research protocol put in place and the theoretical foundations that presided over the elaboration of these investigative choices, the Prémices project seems to us, to provide the keys to understanding the place of digital technology in pedagogical practices, whether in the classroom or in professional training. Considering the fluid identity (de Singly 2003, 2017) of the actors surveyed, and contextualizing the remarks made in a set of escorting discourses and the circulation of trivial knowledge on the subject, allows us to grasp the intra-individual in greater detail, as well as interpersonal tensions that arise when considering the integration of digital technology in teaching–learning practices. The data collected show a great awareness of the responsibility of these social actors, as well as the demands that weigh on them and the projections of expectations towards them.

As we stated in the introduction to this chapter, the Prémices Project clearly has a proactive aim, defining itself as research for action. Thus, our team has ventured to formulate proposals to offer perspectives for action, particularly in the area of initial and in-service teacher training, as well as for ESPÉ trainers.

Throughout the collective discussions during the analysis of the results, we were reminded of this obvious statement by Serge Boimare: "When one fears teaching, it is the whole pedagogy that is disturbed" (Boimare 2012). In fact, the teacher trainees we met during this survey testify to a difficulty in positioning themselves and having self-confidence in a context of threefold fears: fear of the digital object itself, fear of its practices among young people and fear of lesson planning and related teaching–learning situations. We were struck by the almost total absence of the notion of pleasure when these young teachers talk about teaching practices related to digital technology: fears and uncertainties take over. Similarly, this notion of pleasure (to socialize, get information, search for information, communicate), as well as curiosity, is absent from their discourse when they consider digital technology for students in the classroom, with a strong focus on the "dangers" or "risks" of digital technology. Apart from the fact that we know the link between pleasure and motivation in paying attention – or even committing ourselves – to learning (Galand 2006), this absence seems to us to be all the more problematic as it is at the origin of a strong rift between the systems of values and references of digital activity between the non-formal learning sphere and the formal sphere (Cordier 2019).

Beyond the perception and practices of digital technology in the classroom by teacher trainees, our project has attempted to capture the conception and place of digital technology in their professional training. The results of the quantitative and qualitative surveys give rise to four affirmed positions on our part. Firstly, it seems problematic to us that "digital technology" is isolated in the training models, and therefore, in the understanding of trainers and young teachers alike: "digital technology" corresponds to a teaching unit, to a specific time of the course, most often outside of disciplinary didactic questions. Thus, our second point is as follows: we urge a cultural approach to digital technology in professional training. It is a question of trivializing its use, not by denying its specificities (in terms of functions as well as of skills mobilized by its use), but considering that it is naturally part of a potential teaching and learning situation. If we question a lot, and rightly so, the relevance of the use of digital technology in the classroom, we should in fact do the same for the use of audiovisuals, images and even texts, as pedagogical levers for teaching, and as teaching and learning objects. Thirdly, there is an unclear distinction in professional training between teaching with digital technology and teaching digital technology. While pedagogical tools are absolutely necessary to support the possibilities of planning lessons and deploying a training ecosystem in the classroom, digital training cannot be limited to this instrumental aspect. It is important that the choices of tools be discussed, that they be anchored in projections of teaching–learning situations and that they be the object of tenuous links with disciplinary didactics. Using digital technology in the classroom seems all the more risky for young teachers because, as we have seen, they have little sense of technical and conceptual expertise. It is therefore important that each teacher be a repository of an information culture, made up of an understanding of the social, cultural, economic and political stakes of digital objects. The fourth and final point we wish to emphasize, in light of the results of this research, is the need for ongoing training for the trainers themselves, whose representations of teaching authority, potentially undermined by the digital world[13] and

13 The numerous reactions to the Tribune published in Libération on September 18, 2018 on the use of computers by students during university courses have highlighted this fear; the subject has provoked, and exacerbated in equal measure, exchanges among instructors in Espé. Esteves, O. (2018). Ordinateur à l'université : combien y'a-t-il d'étudiants dont on ne voit

pedagogical practices, may sometimes seem out of step with the needs of young teachers. In order for the initial and ongoing training of teachers in primary and secondary education to be truly efficient, we call for an ambitious policy of lifelong training for trainers. Since the introduction of the master's in teacher training courses in 2013, the Ministry of Education has created an online space significantly entitled "Ambition: Teaching"; our dearest wish is that a subspace be created for teacher trainers in INPÉ: "Ambition: Training for Teaching".

3.5. References

Baltz, C. (1998). Une culture pour la société de l'information ? Position théorique, définition, enjeux. *Documentaliste – Sciences de l'information*, 2(35), 75–82.

Baltz, C. (2015). Cyberculture. In *Culture informationnelle : vers une propédeutique du numérique*, Ihadjadene, M., Saemmer, A., Baltz, C. (eds). Hermann, Paris.

Beck, U. (2001). *La société du risque : sur la voie d'une autre modernité*. Flammarion, Paris.

Boimare, S. (2012). *La peur d'enseigner*. Dunod, Paris.

Capelle, C., Cordier, A., Lehmans, A. (2018). Usages numériques en éducation : l'influence de la perception des risques par les enseignants. *Revue française des sciences de l'information et de la communication*, 15 [Online]. Available at: https://journals.openedition.org/rfsic/5011 [Accessed 1 January 2021].

Charaudeau, P. (2004). Tiers, où es-tu ? À propos du tiers du discours. In *La voix cachée du tiers : des non-dits dans le discours*, Charaudeau, P. and Montes, R. (eds). L'Harmattan, Paris.

Cordier, A. (2010). Face à un objet d'enseignement-apprentissage technologique : la reconfiguration des interactions enseignants-enseignés. Speech, Colloque scientifique Ludovia, Ax-Les-Thermes [Online]. Available at: http://www.ludovia.com/news-103-695.html [Accessed 1 January 2021].

Cordier, A. (2011). Imaginaires, représentations, pratiques formelles et non formelles de la recherche d'information sur Internet : le cas d'élèves de 6ème et de professeurs documentalistes. PhD Thesis, Lille 3, Lille.

jamais les yeux ? *Libération.fr* [Online]. Available at: https://www.liberation.fr/debats/2018/09/18/ordinateur-a-l-universite-combien-y-a-t-il-d-etudiants-dont-on-ne-voit-jamais-les-yeux_1679549.

Cordier, A. (2012). Et si on enseignait l'incertitude pour construire une culture de l'information ? *Communication & Organisation*, 42 [Online]. Available at: http://archivesic.ccsd.cnrs.fr/docs/00/80/30/91/PDF/CORDIER_Et_si_on_enseignait_l_incertitude.pdf [Accessed 1 January 2021].

Cordier, A. (2017). Les enseignants, des êtres sociaux pris dans des injonctions paradoxales. *Hermès*, 78 [Online]. Available at: https://halshs.archives-ouvertes.fr/halshs-01598221/document [Accessed 1 January 2021].

Cordier, A. (2019). Ados en quête d'infos : de la jungle à la steppe, cheminer en conscience. In *Accompagner les ados à l'ère du numérique*, Lachance, J. (ed.). Presses Universitaires de Laval, Laval.

Cordier, A. and Liquète, V. (2014). La culture générale face à l'information. In *Cultures de l'information*, Liquète, V. (ed.). Hermès – Les Essentiels, Paris.

Dechaux, J.-H. (2015). Intégrer l'émotion à l'analyse sociologique de l'action. *Terrains/Théories*, 02 [Online]. Available at: http://teth.revues.org/208 [Accessed 1 January 2021].

Dioni, C. (2008). Métier d'élève, métier d'enseignant à l'ère numérique : rapport de recherche pour l'INRP [Online]. Available at: https://edutice.archives-ouvertes.fr/edutice-00259563/document [Accessed 1 January 2021].

Dubar, S. (1991). *La socialisation : construction des identités sociales et professionnelles*. Armand Colin, Paris.

Galand, B. (2006). La motivation scolaire : approches récentes et perspectives pratiques. *Revue Française de Pédagogie*, 155 [Online]. Available at: https://journals.openedition.org/rfp/56 [Accessed 1 January 2021].

Lehmans, A., Capelle, C., Liquète, V. (2017). Ce que le numérique change aux concours de l'éducation nationale. *Hermès La Revue*, 78, 189.

Perrenoud, P. (1999). *Enseigner : agir dans l'urgence, décider dans l'incertitude*. ESF, Paris.

Postic, M. and de Ketele, J.-M. (1988). *Observer les situations éducatives*. Presses Universitaires de France, Paris.

Reuter, Y. (2007). La conscience disciplinaire : présentation d'un concept. *Éducation et Didactique*, 1(2) [Online]. Available at: https://journals.openedition.org/educationdidactique/175 [Accessed 1 January 2021].

Rinaudo, J.-L. and Ohana, D. (2009). Entre aise et malaise. In *Environnements numériques en milieu scolaire*, Rinaudo, J-L. and Poyet, F. (eds). INRP, Lyon.

Rinaudo, J.-L. and Poyet, F. (1999). *Environnements numériques en milieu scolaire : quels usages et quelles pratiques ?* INRP, Lyon.

Sallaberry, J.-C. (1996). *Dynamique des représentations dans la formation*. L'Harmattan, Paris.

Sartre, J.-P. (1985). *Critique de la Raison dialectique, tome 1 : théorie des ensembles pratiques*. Gallimard, Paris.

de Singly, F. (2003). *Les uns avec les autres : quand l'individualisme crée du lien*. Armand Colin, Paris.

de Singly, F. (2017). *Double je : identité personnelle et identité statutaire*. Armand Colin, Paris.

Veyrie, N. (2014). Quelle pédagogie pour quelle prise de risque ? *Le Sociographe*, (45), 73–81.

PART 2

Risks in the Light of Socio-Economic Issues

4

Top Managers Confronted with Information Risks: An Exploratory Study within the Telecommunications Sector

4.1. Introduction

The digital transformation implemented through numerous projects is the reality of all industries and organizations, public or private. The use and integration of new digital technologies allows them to find levers for improvement (processes, customer relations, etc.) and innovation (products, services, business model) in order to create value and face competition (Matt *et al.* 2015). However, going digital also involves a number of risks requiring organizations to anticipate change and protect themselves from potential harm (Stewart and Jürjens 2017). Digital risk is a consubstantial topic in stakeholders' practices (Capelle 2018). It is of widespread concern for everyone, either in their private or professional lives. As for organizations, they are becoming more attentive to the challenges posed by digital technology and the growing number of risks (financial risks, computer breaches, cyberattacks, malicious attacks on their information assets, etc.) to which they are exposed. These issues fall within the scope of digital security risk management in the digital environment. The Organisation for Economic Co-operation and Development notes that the focus previously placed on the security of information systems and networks is now broadening to encompass the security of economic and social activities that rely on new digital technologies (OECD 2015). From this

Chapter written by Dijana LEKIC, Anna LEZON-RIVIÈRE and Madjid IHADJADENE.

point of view, risk involves aspects related to the physical and digital environment, the people involved in the activities and the organizational processes.

In order to better control multiple risks and, in particular, those related to digital technologies, organizations are adopting a comprehensive risk management approach as an integral part of their decision-making process (Dagorn and Poussing 2012; OECD 2015). As a result, top managers play a major role in deploying this approach and raising awareness among their employees (Munir *et al.* 2017). Their practices, commitments and concerns in this area are of great importance to the success of the organizations' risk management approach in this hybrid environment, which is both digital and physical. However, top managers' behaviors and their perception of information risks have not been studied very much.

We are interested in information risk from the perspective of top managers. Our research work, presented in this text, is part of a qualitative empirical study on the information behavior of a group of top managers. It seeks to answer the following questions: What role do top managers play in the overall information risk management process? What are the information risks perceived by these actors, how are they managed and what are the possible difficulties in applying risk management measures? To answer these questions, we shall rely on the results of our study conducted with 22 telecom network unit managers within a large French group.

4.2. Information risk: the conceptual field

Definitions of "risk" are numerous and differ depending on the context in which it is observed. The common thread in existing definitions is the association of risk with the possibility of negative consequences (Léger 2013). David Le Breton (2017) highlights the polemical aspect of the concept, stating that this issue of risk applies to various fields: legal, economic, ethical, political, social, etc., each of which has its own interpretation of risk (Lemieux 2010). Hence, it is necessary to associate risk "with a context and with a system through which we can give it meaning", to avoid confusion and imprecision (Sarrasin 2004).

Key concept	Domain	Definition
Information risk	Security and information technology	Information risk is the possibility of harm, negative consequences, or undesirable results (outcomes) associated with the selection, shaping, transfer, and use of information (Léger 2013).
		Information risk is thus associated with the process of information processing, which takes place as a component of the information system. Based on this association, one might be tempted to consider information risk as the probability that a threat will exploit the flaws in information systems by negatively impacting them (Vallès 2015).
	Law	Information risk is understood according to the litigation handled and grouped into two categories: the risk of seeing information consulted against the will of its holder and the dissemination of false or misleading information, voluntarily or not (du Manoir de Juaye 2014).
	Economic intelligence	Information risk is the manifestation of information, whether proven or not, that is likely to modify or influence the image, behaviour or strategy of an actor. Its impact can result in financial, technological or commercial losses (Harbulot 2005).
		Information risks do not only concern the management of weaknesses in the computerized information system, but also the more general risk of the organization (Goria and Afolabi 2007).
Info-documentary based risk	Records management/ Archiving	Risk related to records will arise from the possibility that a threat will exploit a vulnerability in the management of business records or in the systems and processes used to create, communicate, update, or destroy them (Lemieux and Krumwied 2011).
		Two major risks, from the point of view of information management, can be schematically identified: long-term conservation, a risk linked to the obsolescence of media and formats, but also existing organizations; identification and traceability, in connection with what some call "infobesity" and the legal risks inherent in the management of the mass of data and documents produced or received by organizations within the framework of their activity (Goubin 2016).

Key concept	Domain	Definition
Digital risk	Risk management and security	Digital risk includes: the impact of natural disasters on data centres and communications infrastructure, system failures, acts of criminal intent to steal online banking information or extortion (cybercrime) as well as corporate and nation-state sponsored espionage to steal intellectual property. It also includes penetration or disruption of a country's IT infrastructure (cyberwarfare), online terrorist activities (cyberterrorism) and militant groups using the Internet to achieve their goals (cyberactivism) (Lloyd's 2010).
	Economy	"Digital security risk" refers to a category of risks associated with the use, development, and management of the digital environment in any business (OECD 2015).
Cyber risk	Risk management	Cyber risks are the consequences of a breach of digital data held and/or managed by the company, whether owned by the company or entrusted to it by third parties, as well as the consequences of a computer system breach (ANIA 2015).
	Security	A cyberattack is an attack on computer systems carried out with malicious intent. It targets various IT devices: computers or servers, isolated or in networks, connected or not connected to the Internet, peripheral equipment such as printers, or communication devices such as cell phones, smartphones or tablets. There are four types of cyber risks with various consequences, directly or indirectly affecting individuals, administrations and companies: cybercrime, image damage, espionage, and sabotage (Government of the French Republic, 2019).
Computer risk	Computer science	"IT risk" (or "information and communication technology – ICT – risk," or "information system risk") corresponds to the risk of loss resulting from the inadequate organization, malfunction, or insufficient security of the information system, understood as the entirety of the system, network equipment and human resources intended for the institution's information processing (Andriec *et al.* 2018).

Table 4.1. *Examples of risk definitions applicable to information*

Based on a selection of writings dealing with the issue of risk in relation to information and the digital environment, we also note the lack of unanimity regarding the term used and its definition. Table 4.1 lists some of the most frequently mentioned concepts: information risk, computer risk, digital risk, digital security risk, cyber risk and the risk we call "info-documentary". The definitions proposed or the details of the concept used by the authors highlight the diversity of meaning according to the fields, be it security, economic intelligence, IT, economics, law, archiving or records management[1].

In the definitions for information risk, the focus is on the risks related to the information itself, that is, the content and not the container. In this regard, Marc-André Léger (2013) argues that it is information, not IT, that contributes to the creation of a competitive advantage for an organization. As a result, the attention of organizations must be focused on information resources. The goal of managing these risks is to prevent "the unauthorized, intentional or accidental disclosure, transfer, modification or destruction of information held by an organization" (*ibid.* 2013, p. 11). Information risk is also considered through litigation and infringement of the rights of the organization or individuals. In the field of law, this reflection includes damage to reputation, a notion that we also find in the field of economic intelligence. The information risk is observed in a broader sense. It is not limited to the organization's information but takes into account all information (internal and external) that can be used in an image creation/destruction process. According to this approach, information risk management aims to identify weaknesses in the organization's overall information system, not just those in the computerized system.

In the field of information and records management, the challenge is to guarantee the durability of information and business documents, their identification, authentication and validity for the duration of their administrative and/or legal usefulness. The durability of information is related to the durability of its mediums, thus requiring a certain knowledge of IT risks. Managing information-documentary risks makes it possible to

1 Fournier *et al.* (2005) define records management as "a function of organization and management that applies to documents, data and information, whatever their form and medium, produced or received by any public or private organization in the course of its activities".

ensure the rights of the organization and/or the actors and the continuity of the activities.

Generally speaking, in the definitions of digital risk, cyber risk and computer risk, the focus is on the digital environment. In these reflections, information risk is only partially covered, since information on traditional media is not taken into account. IT risk focuses mainly on threats that can affect the technical infrastructure (hardware, software and network) and thus compromise the organization's information and its activities. Cyber risk refers to attacks in the digital space that can impact computer systems and digital data at the level not only of organizations but also of the state and its citizens.

Having examined the definitions cited, we can adopt Bruno Gruselle's (2013) consideration, which states that information risk "goes beyond the threshold of digital risk" since it also covers information on physical media. Moreover, we should not neglect "oral" information. Our approach to information risk draws on the thinking and work of Victoria Lemieux and Ember Krumwied (2011), Marc-André Léger (2013), Lyonel Vallès (2015) and Emilie Goubin (2016). We draw on the definitions of information risk and info-documentary risk.

Thus, for the purposes of our research work, we define information risk as the possibility of negative consequences or undesirable outcomes (impacting the organization or its actors) associated with the processing, management or use of information (any form and format included) in the digital and physical space.

This work is situated in the context of the digital transformation of the company Orange and focuses on the information risks perceived by the managers of the telecom network in the digital space. We have indeed favored a holistic approach to information risk, taking into account the evolutions of the organizational and informational environment. However, as our results will confirm, information risk cannot be considered solely in the digital space and can, depending on the approach adopted in an organization, include IT risk or vice versa.

4.3. Controlling information risks: Security policy

In order to control the risks of all kinds that they have to face, organizations adopt a security policy and an associated risk management approach. The content and structure of these approaches vary from one organization to another.

Within the Orange Group (Figure 4.1), where we conducted our study, a document entitled General Security Policy (GSP) sets the security principles and rules regarding information, physical assets, people and the environment (Orange 2017).

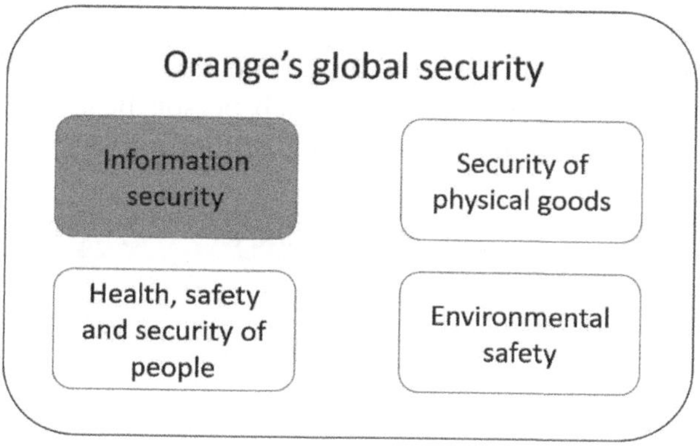

Figure 4.1. *Key elements of security within Orange (adapted from Orange (2017))*

Information security occupies a central place in the security policy. Information risks are assessed in relation to the criteria of confidentiality, integrity, availability, accountability and non-repudiation of information. By means of a risk and threat analysis, information security needs and measures are determined for each entity, as well as scenarios and strategies for managing the identified risks. All risks are assessed and then controlled according to the impact they may have on the organization (vital, critical, sensitive and zero or near-zero impact). The objective of the measures put in place is to protect against disclosure, alteration, destruction, unavailability, usurpation, repudiation, data theft and misinformation. In addition, information protection measures must comply with internal and external requirements (current regulations, contractual clauses, national guidelines, etc.). Within the organization, three types of information security measures are distinguished: technical, procedural and organizational.

The GSP specifies that, regardless of the medium, information must be protected throughout its life cycle, from its creation to its destruction or archiving. It emphasizes that all measures related to the management of information by entities or actors must be in compliance with the Group Retention and Archiving Policy.

The security approach is part of a continuous improvement process, which also includes a security culture. Security is presented as the concern of all internal (employees, trainees, post-graduates, interns) and external (customers, service providers, partners, subcontractors, etc.) stakeholders involved in the organization's activities. The approach involves making everyone responsible (Figure 4.2), with each person bearing their share of responsibility and contributing through their actions to risk control.

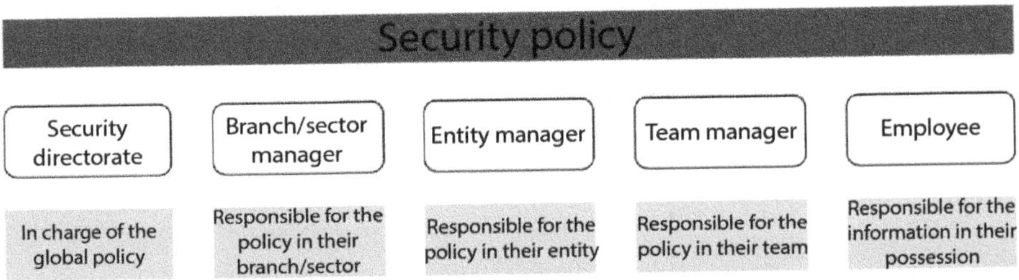

Figure 4.2. *Shared responsibility in the approach to security (adapted from Orange (2017))*

The security directorate is responsible for the company's overarching security policy. It supports the entities in their security initiatives, coordinates security actions, makes recommendations and assists with crisis management. As regards the entities, each manager is responsible for security and its deployment within their own entity. They appoint security managers for each area (security of physical assets, information, the environment and people) and allocate the resources needed to set up a security management system in their entity. They also ensure that the employees of their entity adhere to the security approach, are aware of the risks and the legal and regulatory obligations, as well as the various risk management measures deployed within the organization. The team manager is responsible for security and the application of the measures planned within their team. Finally, each employee is responsible for the information handled or in their possession. They must respect the defined rules and keep themselves informed of the risk management process and measures.

The risk management approach also includes information security awareness actions. The aim is to develop employees' knowledge of information security and thus make them more responsible in this respect. The policy recommends that all stakeholders should integrate the approach as soon as they join the organization and have the necessary knowledge (security measures, obligations, good reflexes, etc.). The obligation of the entity's manager is to implement awareness actions for their employees. These must be updated and conducted regularly in order to remind employees of their security reflexes and to broaden their knowledge in this area.

The life cycle of information and communication technologies is governed by taking into account security needs. Any development of technologies must be done according to the predefined method (expression of need, development cycle, operation, end of service life) (Orange 2017). In connection with these technologies, an access management policy makes it possible to manage users' rights according to their needs (function, professional activity, movements within the organization, etc.) and, thus, to guarantee the integrity and confidentiality of information.

This review of the reference texts presents the strategy adopted by the organization in terms of information risks and applied information security. It allows us to situate the context of our study, as well as the elements of the discourse of the group of actors studied, and to understand how this approach is carried out at the level of the group's entities. Capturing the security requirements and measures adopted at the highest level of the organization serves as a basis for studying the security issues, difficulties and practices of the actors in a particular domain. We have approached them from the perspective of the leaders of the telecom network entities.

4.4. Information risk and management

Numerous scientific works highlight the role of the top manager in information risk management and their influence on employee behavior.

Previous studies and information security experts present the involvement of top managers as a determining element for the effectiveness of the information risk management approach (Ghernaouti and Aghroum 2012; Ponemon Institute 2016). Paul Williams (2007) states that their role is to ensure that the information security policy is understood and adhered to. Aside from a company's top managers, entity or business domain leaders play a central role. Their domain knowledge allows them to identify sensitive and confidential information, to control security-related costs, avoiding excesses, and to allocate access to information appropriately. Quing Hu *et al.* (2012) showed that the involvement of senior management in information security actions has a direct impact on the compliance of employee practices with security policies. Jean-François Berthevas (2013) emphasized the importance of their involvement in the development of the information security policy as well as in its promotion through various means of awareness. Detmar Straub and Richard Welke (1998) postulate that risk can be managed or reduced when managers are aware of existing security measures and implement the most effective controls. These authors conducted a study on the level and extent of managers' knowledge in two American data processing and marketing companies. They showed a lack of knowledge of certain measures and actions to reduce information risks (e.g. deterrence methods, systematic and targeted risk detection) among the top managers of these two companies. Likewise, there was very little awareness among the top managers of the actions to be taken to remedy the damage caused.

Similar results were obtained in a more recent study by Csaba Kollár and Jozsef Poór (2016) focusing on information security aspects in the context of the digital workspace. The authors were able to find a lack of awareness of new information risks among Hungarian top managers. Furthermore, their analysis revealed that some of the top managers interviewed were not even aware of the risks in connection with certain digital domains and digital services mentioned in the questionnaire (e.g. the cloud, augmented reality, social media, etc.). In their answers to the open-ended questions posed, security was mentioned mostly in relation to data, networks, computer systems, access codes and antivirus protection. The human aspect did not figure among the answers obtained. Jean-Noël Ezingeard and Monica Bowen-Schrire (2017), however, sought to identify the factors motivating organizations to review and update their security policy. In particular, they noted that an awareness of risks by top managers influences the frequency of these revisions and their results. Finally, a literature review by Zahoor

Ahmed Soomro *et al.* (2016) showed that top managers are involved in the risk management process through many activities (planning and deploying measures, training, monitoring behaviors, etc.). This is how they contribute to the quality of risk management and information security.

4.5. Study methodology and the stakeholder group

Our empirical study was conducted as part of a research project within Orange, the French telecommunications company. The study focused on the top managers of the entities of the telecom network domain. Rooted in constructivism, the research methodology consisted of a circular and iterative approach to data collection and analysis (Guillemette and Luckerhoff 2012). During the collection phase, we drew on multiple resources. First, business documentation was identified and gathered from the organization's internal devices (intranets, collaborative spaces, corporate social network). The analysis of this documentation allowed us to acquire knowledge on this domain and prepare the survey. We then conducted telephone interviews according to Brenda Dervin's (2008) sense-making methodology.

The study sample consisted of the top managers of the intervention units (IU) and network steering units (NSU)[2]. In total, we interviewed 22 out of 27 top managers (the target population), comprising 18 IU managers and 4 NSU managers. These managers had several years of experience in the network field and held high-level responsibilities. Their job consisted, among other things, of deploying the company's strategy within the entity and organizing and leading the construction and maintenance activities of the telecom network infrastructure. The interviews were conducted in narrative form. Each exchange began with a short presentation of the entity and the functions of the manager. It continued by asking the manager to recall a recent work situation where they had been confronted with a lack of information. We asked additional questions to collect data that would allow us to gain a deeper understanding of the situation and elements related to its context. With the consent of those involved, all the interviews, which lasted

2 These entities bring together all the professions involved in the construction and maintenance of the company's telecom networks. They have operational and functional structures (human resources, management control, communication) and have from 300 to more than 1,000 employees.

an average of 60 minutes, were recorded for the purpose of analysis. The data thus collected was transcribed and made anonymous.

All the collected data was analyzed according to the sense-making and grounded theory methodologies. The researcher examined the behavior and actions taken by the actor along several dimensions: cognitive (thinking, thoughts, moments of constructing meaning), affective (emotions) and situational (a need, lack of or search for information strategies; results and use of information). Analysis of the data according to the concepts of sense-making (Dervin 1983) was done manually. In parallel, the data set was studied by classifying the concepts according to the grounded theory in the NVivo tool. This twofold analysis process allowed us to go further in the use of the collected data set.

4.6. Information risk: The perspective of top telecoms managers

The analysis of the research data allowed us to understand the issues and constraints related to information risks and information security for the telecom network professions and, more specifically, the top managers of the network units. Below, we present the results obtained concerning the role of this group of actors in the information risk management process, in the deployment of information security measures adopted by the organization and the impact of these measures on the availability and access to information.

4.6.1. *Top managers as responsible for information risk management*

Considered as the intangible capital of organizations, information has become one of the concerns of managers and a subject of major importance. The increased use of digital technologies in the realization of activities has exposed manipulated information to new risks that can impact the entire organization. With the changes within and outside the organization (increased competition, regulatory requirements, etc.), information has gained the attention of all actors and, in particular, of top managers.

> Information has become the focus of our attention (IU manager, interview 09[3]).

According to the following top manager, digital technology has had a strong awe-inspiring effect, sometimes diverting attention from an increase in information and digital risks. To prevent damage and ensure information security, he believes that these risks must be analyzed and controlled.

> For the moment, I think we're still a bit amazed by the capabilities of the tools. And we're going to be more and more confronted with potential issues of hacking or incomplete data or customer data that we need to measure and control (IU manager, interview 16).

Consideration of information risks has become essential for any digital project. Referring to the projects of the evolution of the information system and the technologies used, the following top manager indicated that solicitation concerning data protection were more and more common.

> We are being solicited more and more, and rightly so, with regard to sensitive data and customer data […] (IU manager, interview 08).

With high stakes for the organization (brand image, customer confidence, financial losses, etc.), digital projects must include the necessary measures to secure information. Whatever the system deployed and used by the actors, the main concern of the manager is that it is secure. The subject was raised in connection with the storage of data "in good conditions" and on secure servers.

One of the managers indicated that it was a case of "advancing the use of tools" (IU manager, interview 16). This new digital context and the issues it entails make it possible to frame uses and drive all the stakeholders of an organization to become aware of the risks involved.

3 All data that could lead to the identification of the actors interviewed were made anonymous. For this reason, we use the following marking "IU/NSU manager, interview X" corresponding to the function of each actor (manager), the entity managed – IU (intervention unit) or NSU (network steering unit) – and the interview number (X).

Within their respective entities, executives assume responsibility for information security. Due to the daily processing and handling of sensitive data by network stakeholders, information risk management is considered a "critical issue". At the time of taking office, top managers commit, among other things, to protect all information, demonstrating particular vigilance when it comes to confidential and personal data.

> I am responsible, of course, for the security of the data since we are handling sensitive data, we are handling customer data. That means we have their address and their phone number. And so, we must be extremely vigilant (IU manager, interview 10).

> So, I signed something, not too long ago, which says that I am responsible for the security of the data and the manipulation of the data which takes place in my unit, to prevent it from falling into someone else's hands… And I have to do everything possible to prevent… to prevent personal information falling into someone else's hands (IU manager, interview 07).

The role of the manager is to deploy the company's security policy and measures in their entity. In addition, they ensure compliance with all requirements (laws, regulations, national directives) that apply to the company's activities or, more specifically, to the business domain. For example, network entities must comply with the requirements of the telecommunications regulatory authority (ARCEP[4]).

> There is a subject that is being discussed more and more and, well, we are still in the world of telecoms under an authority called the ARCEP. We have to be very careful with any data we use [...] (IU manager, interview 16).

The top manager also has an obligation to ensure that the information security measures are appropriate in a business context and for the requirements in force. When this is no longer the case, they must take the necessary actions to adapt them. A recent example concerns the new General

4 The *Autorité de régulation des communications électroniques et des postes* (ARCEP) is an independent administrative authority responsible for regulating the electronic communications markets.

Data Protection Regulation (GDPR). For top managers, this regulation has resulted in the development of an action plan and new compliance measures.

These results show that risk management and information security are part of the roles and activities of executives. The executive is the one who carries the company's strategy in this area and its values. They ensure the deployment of information protection measures, their evolution, as well as their respect of the actors for their entity. For the top manager, the control of information risks represents a current problem requiring important efforts. The main issue is to maintain the agility of work with the digital devices deployed, while keeping the information and data secure. This is a subject that focuses the attention of the managers and requires their involvement. Their role is to find the best solution at the given time. To achieve this, they must rely on their knowledge of the domain and reference documents, as well as experts in information security, networks and information systems.

4.6.2. *Information risk management*

The field of the network includes frequent work in the field. Consequently, the actors are equipped with portable digital devices and remote access to the information system. This is also true for the managers of the network units, who are very often on the move: "I am almost never at my desk" (NSU manager, interview 22). Given these conditions, the information they have at their disposal may be subject to malicious acts. Thus, among the main risks perceived by telecom executives are the loss or theft of information, the risk of hacking and risky behavior.

The management of these risks primarily implies the implementation of technical measures. The challenge is to prevent the loss of information due to technical problems. It is also a question of ensuring the protection of information in the event of theft or loss of the working device.

> We have a lot of laptops, obviously with the risk of misplacing them, getting stolen or whatever (IU manager, interview 12).

In addition to these various causes of information loss, the digital context involves the risk of hacking and therefore of data leakage. The possibilities of hacking are numerous "in the digital world", and this risk is considered a "major concern" that must be managed and "put under control".

The last eventuality mentioned by managers is risky behavior. Certain behaviors of the actors can, in fact, lead to involuntary damage and harm the company. In the course of their activities, the security reflex is not necessarily present.

> I'm not convinced that, in the heat of operational action, we think about that (IU manager, interview 16).

This last actor drew our attention to the fact that employee practices are also evolving in this new digital context. Certain practices are developing among them, without management being involved or informed. The manager cited the example of information sharing between employees via instant messaging on the app WhatsApp. This practice was presented to management at a board meeting. One of the employees explained that this device allowed field workers to exchange information throughout the day, to ask for advice, for help with problems encountered, or simply to arrange to meet for a lunch break. The use of external devices is not recommended, as the risks of hacking, theft or data leakage are higher. The first reflex of managers in such a case concerned data security:

> Is the data we transmit on WhatsApp secure? (IU manager, interview 16).

The executive cites an example of the technical level where the use of this device to exchange information can have strong impacts:

> If I'm doing a router configuration at a customer's location and that customer is a bank, if it gets hacked, you can imagine what that can do (IU manager, interview 16).

This type of behavior opens the door to threats such as data disclosure and theft. In order to control these threats, awareness and vigilance are required whenever information or work devices are used outside the secure area.

The right approach is to prioritize and promote internal digital devices configured in accordance with security measures and recommendations for employees. In addition to technical measures, managers have identified measures to raise awareness, train and remind their employees of the risks involved. The challenge was to change habits and evolve existing usage, as one of the managers pointed out. In this specific case, the solution was therefore to inform and explain to the actors the risks of the practices developed and also to propose alternatives to them. The idea was always to replace the non-compliant device with an equally practical internal alternative (e.g. Skype Enterprise).

In relation to practices and uses, the managers mentioned in particular the younger generations of employees. They have a "certain skill" and a stronger reflex in their use of digital technology. Sharing information in different communities via social networks or another digital device is quite natural for them. These practices have attracted the attention of managers who feel it is important to establish rules for the use of information.

> This is what we remind people when we recruit technicians and when we have students on placement with us, which is that there are rights and duties in terms of the use of information [...] (IU manager, interview 16).

Technical measures have been seen as a necessary first step in securing information. However, the issue of information security goes beyond the capabilities of technological solutions and requires the implementation of other measures. One executive spoke of the need for a "well-structured framework" with clear procedures and rules. On the one hand, this translates into controlling and managing access to information to ensure that "only authorized people can access the right data" (IU manager, interview 11). On the other hand, this manager believes that it is a matter of defining common rules for the management and conservation of information for all actors. In practice, this is not always the case. Problems linked to the format, naming and loss of information when employees leave, and paper archives that are not properly managed, are among the examples cited by the manager.

Thus, in addition to technical measures, organizational and procedural measures are considered important but are missing. The management of information according to shared rules and in spaces determined by all the actors makes it possible to identify it, to be able to access it and, ultimately, to protect it. This also involves the behavioral aspect and work on information management practices.

To develop security behaviors, awareness is seen as a crucial element in managing information risks. Rules and guidelines are a good foundation. However, once read, they are quickly forgotten. For this reason, an ongoing awareness process was seen as the key to success. The leaders specified that this approach must be applied to all stakeholders. New employees must integrate the approach as soon as they arrive and understand that there are "rights and duties in terms of information use". This will help develop sustainable security behaviors. In addition to awareness-raising measures, the main support for managers in the information risk management process consists of technical measures (encryption, controlled identification and access, etc.). These technical "locks" are considered necessary to avoid "being overwhelmed by innovation" (IU manager, interview 16). That said, the behavioral aspect of information security must be supported by technical measures, which are essential to the information risk management approach and the protection of information.

4.6.3. *Operational challenges related to the information risk management approach*

Information risk management implies that access to the information system is controlled and reserved for authorized persons. Moreover, it is not possible to access information remotely if the device used (PC, tablet, cell phone) is not equipped with a satisfactory security system (e.g. a VPN[5]).

> And clearly, it disables access to information systems as soon as they are on devices that are not completely e-office on the secure system and inside the Orange secure zone (IU manager, interview 08).

The implementation of these measures means that, in some cases, security is experienced as a constraint (technical, functional, organizational). As stated in the previous section of this text, the issue of information security is recognized and considered to be of primary importance. However, security requirements influence the organization of work, make procedures and processes more complex, and are perceived to be limiting thc use of all digital capabilities. One executive draws attention to a "state of affairs", noting that security is sometimes a "brake on all things digital" and

5 A virtual private network is a system of securing information which allows the actor to access a network remotely and securely.

insinuating problems stemming from remote access to information. A concrete example concerns the processing of emails and their encryption using a PKI key[6]. To process sensitive or confidential information, the manager must use this protection system. However, this means that the secured information is only accessible under the predefined conditions, forcing the actor to readjust their behavior.

> We are forced to work more and more with secure emails that are encrypted and only visible with a PKI key. I would do 80% of my e-mails from my phone if I could, but today all these e-mails are only readable on a PC (IU manager, interview 08).

Below, a top manager also noted the tightening of access control to applications. He noted that some information is accessible and viewable only through these authentication certificates (PKI).

> And nowadays, it's become even stricter, since certain applications or information are also encrypted, or are only accessible only in PKIs. With systems for certifying the people who consult, which therefore makes it possible to encrypt what is sent, as well as the consultation of data (IU manager, interview 09).

Information security and risk management concerns all stakeholders involved in the activities and handling of information in the organization's information system. Most often, the managers mentioned the problems of access to the information system in the case of outsourced activities and, therefore, those carried out by partners (subcontractors). An explanation for these difficulties was given, which is that the information system was initially built for internal use, which did not pose any confidentiality problems. The challenge for managers, therefore, is to find solutions to enable partners to do the work required while ensuring a satisfactory level of information security. Among other things, this requires changes to the information system to adapt it to the new organizational context.

> However, accessing it as a subcontractor also means accessing information that could be considered confidential or strategic for us. And so, how do we (1) ensure that this access is possible,

6 A public key infrastructure includes certificates for authentication and data encryption services (Orange internal system).

> because the system just doesn't provide for it; it was designed in a world where everything was done with Orange employees, and the idea of a third party doing this wasn't foreseen, so there are changes to be made. And then (2) how we are going to manage the information afterwards, so that they can access what we ask them to do, but no more? (IU manager, interview 18).

In this context, the manager specified that access to information and security of information is an organizational and contractual matter. The relationships, requirements and rules to be respected are spelled out in the partnership contracts. In the field, compliance with regulatory requirements and the confidentiality of information requires partners to have selective and limited access to the information system (IS). In other words, subcontractors cannot remotely consult all the information in Orange's IS.

Due to security requirements and constraints, access to the information system and visibility of information have been assigned in a segmented manner. Therefore, the right balance must be found according to the activities assigned to each partner.

In addition to access to information, this also means limited access to a number of functionalities. For example, in order to perform certain actions in the field, contractors are required to call for technical assistance.

> So, there are some applications which [the subcontractor] has access to, and then there is a break in the digital chain, meaning that he is obliged to go through the telephone if he does not have access to the information. [...] We are constantly confronted with these data security issues (IU manager, interview 08).

Another example that was given to us concerns the reporting of work done and the updating of information in databases. In connection with these problems, the following manager indicated two possible solutions. The one concerning the act of updating information involves engaging the internal actors to complete this work. The second involves consulting the information system via specific accounts on the site, at the Orange premises.

Lastly, a final practice consists of making a maximum amount of information available to these actors through the application serving as an interface between the partner's information system and that of Orange. The next manager explained that these stakeholders need a great deal of technical information, information about the location of the intervention, the customer's subscription, etc. Because of the limited access to the Orange information system, the goal is to provide them with the most complete information possible in the work orders so that they can be more autonomous. Moreover, this reduces costs, the number of calls and the overload of the technical and support services to the field.

To resolve these issues related to access and rights management, the approach to developing the information system has also changed. Below, the manager indicated that IS developments have started to be done in such a way as to allow simpler rights management for all actors directly from the application concerned (without using an interface for subcontractors).

> Today, the logic is different. We develop a single IS for both employees and subcontractors. And if there is the slightest need to restrict rights or security, it is the application itself that, how can I put it, manages the restrictions; we don't have two applications! (IU manager, interview 15).

The topics of access to information and information security are considered complex and far-reaching. Compliance with security requirements and measures is seen as burdensome in some situations. The main problem, as mentioned by managers, concerns restrictions on access to information and its unavailability for reasons of confidentiality or lack of satisfactory security measures. A significant impact is noted with regard to the work and activities of external collaborators (subcontractors). One of the managers told us that about 70% of the activities in the field are outsourced. Security and confidentiality requirements mean that subcontractors have limited access to the information and functions of the Orange information system. In addition, the IT system, consisting of myriad applications and initially designed for company employees, makes it even more difficult to manage and assign access rights. As a result, work organization and procedures, as well as information management, are more complex. Considering the importance of outsourcing in this business area, ensuring access to the necessary information is essential for business continuity. To overcome these problems of information unavailability, managers implement other, less "agile" solutions, sometimes requiring additional efforts and costs. These facts lead them to the conclusion that the evolution of information systems must continue, in order to make information available at all times, in the same way for internal and external collaborators, and in a secure manner.

The result is that information security is becoming a key element of management practice, particularly in the definition of access rights to information.

4.7. Conclusion

In an evolving technological environment, the control of information risks by organizations has become a complex issue that is the subject of many studies (Desroches 2013; Castelo Branco and Bolliger 2018). To build an approach to managing these risks, organizations define a strategy and implement a set of security activities and measures (AFNOR 2013; Castelo Branco and Bolliger 2018). They rely on various standards, national guidelines, recommendations and methods. In its risk management approach, Orange uses, among others, the French EBIOS method published by the *Agence nationale de la sécurité et des systèmes d'information* (ANSSI), the French National Agency for Security and Information Systems. The method offers tools that allow organizations to assess and implement the necessary measures to control digital risks. It includes a five-step or "workshop" approach and can be adapted/applied in any organization, regardless of its size, area of activity or level of information system design (ANSSI 2018). It is an iterative approach that is presented in three levels of risk management: simple, elaborate and advanced.

In addition to technical and organizational measures to manage information risks, acculturation and adherence to the risk management approach by all stakeholders in the organization is required (ICSI 2017). The behavior of humans, and thus the employees of organizations, is still presented today as the main flaw in information security systems (Stewart and Jürjens 2017). Hence, organizations are focusing on developing a security culture. This has been defined as "a set of ways of doing and thinking that are widely shared by actors in an organization focused on controlling the most important risks associated with its activities" (ICSI 2017, p. 9). Through a good organizational culture, the company can influence employee engagement, help them understand information protection issues and devices, and thus promote behaviors consistent with its information security policy (Sikolia 2013).

The management of information risks is a subject of major importance for telecom managers, who are equipped with numerous digital devices. The information risks perceived by this group of actors can be classified according to the categories defined by Karen Loch *et al.* (1992). Risks internal to the organization concern technical problems that can cause the loss or alteration of information. Risks external to the organization involve malicious actions (hacking, theft). Moreover, these two categories of risks (internal and external) also include human behavior.

The results of our study also show that the telecom executive assumes the role of information security manager. This role includes the deployment of the organization's information security policy within their entity, the implementation of information risk management measures and their evolution according to changes in the environment. This result confirms the work of Jean-François Berthevas (2013) on the key role of management at the local level with regard to the organization's strategy and values in terms of information risk management. Thus, the telecom manager organizes activities and human resources in such a way as to guarantee compliance with requirements. They monitor the use of devices and information, raise awareness among their staff and instruct them to behave in accordance with security policies. These activities correspond to the managerial activities related to information security proposed by Paul Williams (2001): policy development according to business needs, the implementation of adopted measures, monitoring, awareness and training, etc. The results of our study contribute to the existing work by confirming that the activity of information risk management is part of the managerial work of telecom managers. We can therefore consider that the role of information security manager is part of the informational roles of managers presented in Henry Mintzberg's (1973) categorization.

In some work situations, the requirements for managing information risks are perceived as constraints (functional, organizational, technical). They modify the behavior of actors and their use of digital devices. Thus, access to and processing of information should be carried out under satisfactory security conditions. This result complements the work of Dijana Lekic and Anna Lezon Rivière (2018), which confirms the influence of security requirements on information sharing in the digital environment. We also find that information overload is perceived by leaders as a form of information risk.

Although information risk from the point of view of top managers has not been explored much in previous studies, our study sheds light on this subject and contributes to enriching knowledge on their information security practices. It allows organizations to identify weaknesses and difficulties in implementing an information risk management approach in an operational domain such as the telecom network domain. Finally, our results highlight the need for employee awareness of the right behavioral reflexes in the face of information risks and for a widely deployed security culture.

This initial exploratory study has two limitations. First, our study focuses on top managers from a single field of activity, that of the telecom network. Second, the security aspect was not the main focus of our research. It is included in a broader study of all the information behaviors of telecom managers. Therefore, a return to the field would allow us to deepen the work already conducted. The information risks would benefit from being studied from the perspective of other actors in the telecom network entities. Furthermore, the study could be continued with managers from other levels of responsibility and business areas.

4.8 Acknowledgments

This study was carried out thanks to the support of Orange, a French telecommunications company. The authors thank all the top managers who participated in the study, as well as the reviewers and editors for their constructive suggestions.

4.9. References

AFNOR (2013). *NF ISO/CEI 27005 : Technologies de l'information, techniques de sécurité, gestion des risques liés à la sécurité de l'information*. AFNOR, Saint-Denis, France.

Andriec, M., Carteau, D., Cornaggia, S., Ginolhac, P., Gruffat, C., Le Maguer, C. (2018). *Le risque informatique : document de réflexion*. ACPR Banque de France, Paris, France.

ANIA (2015). Les Cyber Risques. Report, ANIA, GRAS SAVOYE, FINEX Lignes Financières, Paris, France.

ANSSI (2018). EBIOS Risk Manager : une démarche itérative en 5 ateliers. Agence Nationale de la Sécurité des Systèmes d'Information, Paris, France.

Berthevas, J.-F. (2013). Management des réseaux personnels et de la sécurité de l'information dans une perspective d'innovation : le rôle de la culture organisationnelle. PhD Thesis, Université Aix-Marseille, France.

Capelle, C. (2018). Rapport final de projet de recherche eR!SK – Risques numériques et école 2.0. Report, Laboratoire IMS, Université de Bordeaux, France.

Castelo Branco, G. and Bolliger, M. (2018). Gestion des risques informationnels dans les organisations. Master's thesis, Haute École de Gestion de Genève (HEG-GE), Geneva, Switzerland.

Dagorn, N. and Poussing, N. (2012). Engagement et pratiques des organisations en matière de gouvernance de la sécurité de l'information. *Systèmes d'information & management*, 17(1), 113–143.

Dervin, B. (1983). An overview of sense-making research: Concepts, methods and results. Speech, International Communication Association Annual Meeting, Dallas, Texas, USA.

Dervin, B. (2008). *Interviewing as dialectical practice: Sense-making methodology as exemplar. Speech*, International Association for Media and Communication Research (IAMCR), Stockholm, Sweden.

Desroches, C. (2013). La gestion des risques informationnels dans l'entreprise privée : perspective des gestionnaires de la sécurité. Master's thesis, Faculty of Arts and Sciences, Montreal, Canada.

Ezingeard, J.-N. and Bowen-Schrire, M. (2017). Triggers of change in information security management practices. *Journal of General Management*, 32(4), 53–72.

Fournier, D., Morineau, E., Christophe, L., Droulier, S. (2005). Comprendre et pratiquer le records management : analyse de la norme ISO 15489 au regard des pratiques archivistiques françaises. *Documentaliste – Sciences de l'Information*, 42(2005/2), 106–116.

Ghernaouti, S. and Aghroum, C. (2012). Cyber-résilience, risques et dépendances : pour une nouvelle approche de la cyber-sécurité. *Sécurité et stratégie*, 11(4), 74–83.

Goria, S. and Afolabi, B. (2007). Proposition d'une démarche de questionnements pour modéliser un Système d'Intelligence Économique. *International Journal of Information Sciences for Decision Making*, 31(535), 1–12.

Goubin, E. (2016). Les archivistes face au défi de la dématérialisation. *La Gazette des archives*, 242(2), 149–159.

Gouvernement de la République française (2019). R!sque : prévention des risques majeurs [Online]. Available at: https://www.gouvernement.fr/risques/risques-cyber [Accessed 14 October 2019].

Gruselle, B. (2013). *Enquête sur la sécurité numérique des entreprises*. Fondation pour la Recherche Stratégique, Paris, France.

Guillemette, F. and Luckerhoff, J. (2012). *Méthodologie de la théorisation enracinée : fondements, procédures et usages*. Presses de l'Université du Québec, Quebec, Canada.

Harbulot, C. (2005). *L'entreprise face au risque informationnel. Speech*, Symposium sur la sécurité des technologies de l'information et des communications (SSTIC), Rennes, France.

Hu, Q., Dinev, T., Hart, P., Cooke, D. (2012). Managing employee compliance with information security policies: The critical role of top management and organizational culture. *Decision Sciences*, 43(4), 615–660.

ICSI (2017). La culture de sécurité : comprendre pour agir. Institut pour une Culture de Sécurité Industrielle (ICSI), Toulouse, France.

Kollar, C. and Poor, J. (2016). *Organizations in digital age – Information security aspects of digital workplaces. Speech*, Management, Enterprise and Benchmarking in the 21st Century, Budapest, Hungary.

Le Breton, D. (2017). *Sociologie du risque*, 2nd edition. Presses Universitaires de France, Paris, France.

Léger, M.-A. (2013). *Introduction à la gestion de risque informationnel*. CRHOMA, Quebec, Canada.

Lekic, D. and Lezon Rivière, A. (2018). Pratiques de partage de l'information dans l'environnement numérique : cas des dirigeants du réseau télécom. *AIDAinformazioni*, 36(3–4), 107–128.

Lemieux, V.L. (2010). The records-risk nexus: Exploring the relationship between records and risk. *Records Management Journal*, 20(2), 199–216.

Lemieux, V.L. and Krumwied, E.D. (2011). Managing records risks in global financial institutions. In *Managing Records in Global Financial Markets: Ensuring Compliance and Mitigating Risk*, Coleman, L., Lemieux, V., Stone, R., Yeo, G. (eds). Facet Publishing, London, UK.

Lloyd's (2010). *Managing Digital Risk: Trends, Issues and Implications for Business*. Lloyd's, London, UK.

Loch, K.D., Carr, H.H., Warkentin, M.E. (1992). Threats to information systems: Today's reality, yesterday's understanding. *MIS Quarterly*, 16(2), 173–186.

du Manoir de Juaye, T. (2014). Le risque informationnel au filtre du droit. *Documentaliste – Sciences de l'Information*, 51(3), 37–40.

Matt, C., Hess, T., Benlian, A. (2015). Digital transformation strategies. *Business & Information Systems Engineering*, 57(5), 339–343.

Mintzberg, H. (1973). *Le manager au quotidien : les 10 rôles du cadre*, 12th edition. Eyrolles, Saint-Germain, France.

Munir, R.A., Molok, N.N.A., Talib, S. (2017). *Exploring the factors influencing top management involvement and participation in information security*. Speech, Pacific Asia Conference on Information System (PACIS), Langkawi, Malaysia.

OCDE (2015). *Gestion du Risque de Sécurité Numérique pour la Prospérité Économique et Sociale : Recommandation de l'OCDE et Document d'Accompagnement*. Organisation de Coopération et de Développement Economiques (OCDE), Paris, France.

Orange (2017). Politique de sécurité globale pour le Groupe Orange. Orange, Paris, France.

Ponemon Institute (2016). Closing security gaps to protect corporate data: A study of US and European organisations [Online]. Available at: https://info.varonis.com/hubfs/docs/research_reports/Varonis_Ponemon_2016_Report.pdf [Accessed 15 October 2019].

Sarrasin, B. (2004). Risque politique et tourisme : nouveautés et continuités. *Téoros*, (23–1), 12–22.

Sikolia, D.W. (2013). Toward a theory of employee compliance with information security policies: A grounded theory methodology. PhD thesis, Oklahoma State University, Stillwater, Oklahoma, USA.

Soomro, Z.A., Shah, M.H., Ahmed, J. (2016). Information security management needs more holistic approach: A literature review. *International Journal of Information Management*, 36(2), 215–225.

Stewart, H. and Jurjens, J. (2017). Information security management and the human aspect in organizations. *Information and Computer Security*, 25(5), 494–534.

Straub, D.W. and Welke, R.J. (1998). Coping with systems risk: Security planning models for management decision making. *MIS Quarterly*, 22(4), 441–469.

Valles, L. (2015). Le risque informationnel et l'urgence de le gérer de façon adéquate [Online]. Available at: http://lyonelvalles.com/2015/12/20/le-risque-informationnel-et-lurgence-de-le-gerer-de-facon-adequate/ [Accessed 14 October 2019].

Williams, P. (2001). Information security governance. *Information Security Technical Report*, 6(3), 60–70.

Williams, P. (2007). Executive and board roles in information security. *Network Security*, 2007(8), 11–14.

Zicry, L. (2015). Les Cyber Risques Association Nationale des Industries Alimentaires (ANIA) [Online]. Available at: https://www.ania.net/wp-content/uploads/2015/06/retour-sur-atelier-fraudes-et-cyber-risques-ania-13-04-2015.pdf [Accessed 2 January 2020].

5

Cell Phones and Scamming Risks in Cameroon: Users' Experiences and Socio-Institutional Responses

5.1. Introduction

Cell phones are technological tools that are being used more and more. For Cameroonians, cell phone use has spread rapidly and this tool has been a part of daily life since the 2000s. In 2017, the Telecommunications Regulatory Board (*Agence de Régulation des Télécommunications*, ART) of Cameroon counted 19,706,027 mobile subscribers across all operators (ART 2018). Although this figure does not give a precise indication of the total number of cell phone users – some people have more than one subscription – it is worth noting that cell phones are one of the most widely used communication tools in Cameroon. Like Information and Communication Technologies (ICT), which have long been associated with the socio-economic development of Africa (Chevanaz and Paraschiv 2011; Loukou 2012), the cell phone is much lauded in Cameroon for the socio-economic dynamics associated with its use. Furthermore, with respect to mobile money[1], mobile transactions are gradually becoming commonplace among Cameroonians. Subscribers to this service – provided by both MTN and Orange, the two operators active in this sector – went from 2,074,653 in

Chapter written by Freddy TSOPFACK FOFACK and Abdel Bernazi RENGOU.

1 According to ART (2018), "mobile money is for GSM subscribers to virtually store (electronic money) a monetary value in an account associated with the mobile SIM card and/or perform financial transactions (payments, transfers, etc.) from the cell phone terminal".

2013 to 8,003,252 in 2017 (ART 2018). Moreover, according to the same source, the number of transactions carried out increased from 4,100,000 in 2013 to 106,022,289 in 2017.

Use of mobile money services, beyond the Internet – which is generally at the center of scientific work on cybercrime – is thriving in technology-enabled contexts, which now present the risk of what is termed “cell phone fraud” or, alternatively, “cell phone scams”. This concept, which is a type of cybercrime, is used here to refer to any scam initiated from a cell phone, independently of the Internet, that relies on mobile money services made available by mobile phone operators. Although mobile transactions are gradually gaining momentum in Cameroon, the fact remains that they do encourage cybercriminal activity. Thus, it is important to note that, in this work, risk is considered not only as an event that has not yet taken place (Beck 2008) but also as a fait accompli with the associated mitigation actions making it possible to warn others. Thus, the risk presented by cell phone scams mentioned here has two facets: the perceived risk and the obvious risk. The perceived risk refers to the perception cell phone users have of the risk linked to the use of cell phones, in relation to attempted and/or actual scams to which they are subjected, and which are reported on a daily basis. In this context, Bauer (1960) (cited in Chevanaz and Paraschiv 2011) understands the notion of perceived risk as the perception by the consumer of the harmful consequences that may result from his or her actions. Five dimensions emerge, of which the financial risk appears fundamental in this work as the extortion of money is at the heart of scams. The obvious risk refers to the explicit manifestation of this scam, to the negative (financial) consequences of the use of the mobile money service, that is, to cases where the scammers achieve their aims. Such an understanding of risk comes close to the ontological or realist view, which takes into account “the possibility of the realization of an event and its negative consequences” (Kermisch 2012).

Information disseminated in the media, by telecommunications operators and regulators, as well as some scientific studies (Nji and Ebi 2019; Nkongho 2019), refer to scams encountered daily by cell phone users in general and subscribers to mobile money services in particular. The risk of cell phone scams in the Cameroonian context is rife even in a telecommunications sector where security issues are the focus of the various socio-institutional organizations; these include the Ministry of Posts and

Telecommunications (MINPOSTEL), mobile operators, regulators and civil society organizations, whose actions aim to protect cell phone users. In such a context, any observer would assume that the mobile money service would be as secure as possible; this is far from being the case. The *Centre d'alerte et de réponse aux incidents cybernétiques* (CIRT) of the *Agence Nationale des Technologies de l'Information et de la Communication* (ANTIC) notes that mobile bank account hacking was among the 10 most popular attacks in national cyberspace in 2018 (CIRT 2018).

The risk of cell phone scams mentioned above tests the need for satisfaction and security commonly required by any consumer, and legislated by framework law No. 2011/012 of May 6, 2011 on consumer protection in Cameroon[2]. Here we find a paradox: despite the implementation of provisions intended to contain risk, by protecting the cell phone user, the risk is nevertheless still high. How can we understand such an ambiguity? The use of cell phones requires that users have not only technical but also cognitive skills to avoid being swindled. This state of affairs brings to the fore the issue of supervising cell phone users in a high-risk environment. Thus, how can we assess the governance or supervision of cell phone use in Cameroon? More precisely, what are the mechanisms facilitating the upsurge in cell phone scams that affect users on a daily basis? And what socio-institutional action is being taken to deal with this risk? By subjecting cell phone scams to the scrutiny of sociological analysis, this chapter elucidates the processes at work in the socio-institutional governance of cell phone use. This has been achieved by conducting a critical socio-analysis of the mechanisms by which cell phone users are affected by scams and of the socio-institutional dynamics related to the supervision of the use of this communication tool in Cameroon.

A look at the sociological literature reveals the existence of work on ICT scams from a variety of perspectives. For a long time, research has focused on cybercrime, particularly on Internet fraud, with studies that examine the practices of cyber swindlers in order to analyze the modalities of "hijacking", "poaching" (de Certeau 1990) or appropriation of ICT use. To this end, studies include the phenomena of cybercrime in Benin (Tasso Boni

2 Loi-cadre no. 2011/012 du 6 mai 2011 portant protection du consommateur au Cameroun, available at: http://www.art.cm/sites/default/files/documents/loi_cadre_2011_012_06052011_protect_conso.pdf [Accessed 15 September 2019].

2014), Sakawa in Ghana (Perrot 2015) and cybercrime or "Zarguina[3]" in Cameroon (Abia *et al.* 2010; Rengou 2017). Other studies focus on victims of Internet scams and analyze how Internet users are manipulated (Auray 2012), motivations for alerting police and law enforcement agencies (Reyns and Randa 2015; Cross 2018) and Internet users' responses or actions after falling victim to attacks (Benbouzid and Peaucellier 2016), among others. Other work focuses on cybercrime, specifically e-commerce and mobile money, and conducts an analysis of consumer protection laws to highlight strengths and weaknesses (Nji and Ebi 2019; Nkongho 2019).

All in all, the scientific literature is very often interested in downstream actors, in particular the users or victims, and the "disruptive" actors that are the cybercriminals, while neglecting those upstream, namely the supervisory and regulatory bodies. In terms of the latter, the major scientific contributions include studies on: the "demons" or "jammers"[4] of the net and of the action taken by the telecommunications agency of Côte d'Ivoire (Bogui 2009); the role of information and communication technologies in the diversification of forms of urban banditry and the fight against this in Cameroon (Dekane 2016); and the fear of profiling in the context of the regulation of the use of information and communication technologies in Côte d'Ivoire (Bogui and Atchoua 2016). Moreover, the scamming phenomenon is generally only studied from the point of view of Internet users, thus excluding a large proportion of ICT users who are still exposed to cell phone scams, independently of Internet use, such as users of mobile money services. Faced with this scientific deficit, this chapter studies the risk of cell phone scams that prevails in Cameroon by exploring it using both a bottom-up and top-down approach; thus, we study the (potential) victims through a sociology of socio-institutional governance of mobile telephony and furthermore the disruptive actors are also taken into account. This work thus reveals the processes by which mobile hackers are embedded in the telephone realm of users and the socio-institutional actions of actors from above to deal with them. It is also important to emphasize that the mobile money system, as we understand it, is made up of mobile money users, mobile crooks and socio-institutional organizations.

---

3 Cybercriminals or cyber scammers in the Cameroonian context, especially in the Department of Noun in West Cameroon.

4 Naming of cybercriminals in the Ivorian context.

From a methodological point of view, this research is essentially based on a qualitative approach. In addition to the documentary analysis carried out beforehand, a field survey was carried out through semi-structured interviews with the three categories of actors within the cell phone scam system. First, 15 cell phone users, 10 of whom had experienced attempted scams and 5 of whom had been victims, were interviewed in order to describe their experience of (attempted) mobile scams. These individuals consisted of online commercial sellers and students who conduct mobile money transactions on a daily, weekly or monthly basis; the victims were identified from contacts in our interpersonal networks. Second, two officials from the telecommunication companies MTN and Orange, a mobile money and communication credit operator, an official from MINPOSTEL, ART and ANTIC, and four judicial police officers from the Gendarmerie and the Police were interviewed to analyze the mediation practices at work to ensure the protection of cell phone users. Third, five mobile criminals explained to us the operational mechanisms of their fraudulent actions; they were recruited from the social network of the second author who has been working with them for several years in the context of his research on cybercrime issues. In addition, it should be noted that we proceeded with an observation of usage, in particular by guided viewings of the phones of the interviewed users in order to have the scam messages received, and with an observation of the mobile scammers in (attempted) scam situations.

This chapter is structured in three parts. First, the various methods used in cell phone scams are presented, followed by a look at the socio-institutional mediations at work in the monitoring of the telephone sector to address this risk and third, this chapter questions the effectiveness of these mediations in view of the prevalence of this digital risk.

5.2. Mechanisms behind cell phone scamming in Cameroon: exhibiting credulity

Cell phone scams are implemented using a variety of methods with the intention of extorting money from the targeted user. Scammers make use of several strategies to get closer to their "clients"[5], establishing close links and sometimes even making their acquaintance. This is carefully organized and managed to ensure they achieve their aims.

5 This is jargon used by scammers in Cameroon to refer to their victims.

5.2.1. *Setting the scene*

We use this expression to refer to a strategy that cell phone scammers use to make the situation appear credible by establishing a relationship of trust with the potential victims. One technique is to present oneself as a member of their interpersonal network by mentioning people they know with details about their professions and their locations in order to dispel any surprise and mistrust. This involves the use of a set of signs, symbols and images chosen according to the targeted user, which can be likened to "symbolic violence" and whose aim is to subject the interlocutor to the scam. In this regard, one respondent stated:

> … He called me on the phone… And using my name, he told me that we were classmates. Then he mentioned a friend from my class and claimed to be a mutual friend. He asked me about a lot of people I know… I was confident. That's when he made me a business proposal that would later prove to be a hit (scam).

During this scamming process, everything is done to persuade the victim of the veracity of the business in progress. One after the other, the accomplices of the mobile scammer take turns on the phone pretending to be the business partners. Thus, this scam situation is a well-organized, well-crafted and personalized machine where each person plays a specific role. In some of the situations observed with mobile scammers, the same person plays all the roles by changing voices and phone numbers, either by inserting a toothbrush in their mouth or by blocking their nostrils with their fingers. Others use voice changing applications by choosing the one that corresponds to the personality announced to the "client".

Another technique is to instill fear and compassion in the target user. It involves informing a person that a relative of theirs has been the victim of a traffic accident and is being kept in the emergency department of a hospital waiting for money to be sent for the actual treatment, which must be sent as soon as possible so that the relative can be saved. Sometimes the scammers call the alleged victim beforehand, pretending to be a customer service department which is in the process of resetting certain services; the victim is asked to validate the codes they are given and to turn off the phone for a certain period of time in order to confirm the update. In addition, others make calls with masked numbers, pretending to be relatives of the targeted user who are stuck in a customs clearance operation, whom they would have liked to surprise on arrival without informing them beforehand, and who have a very tight deadline; they then ask the victim to deposit a sum of money via the telephone number they are using to communicate. The personalization of the scam scene thus proceeds through a "*mise en scène*" (Goffman 1974) insofar as the scammers consciously play roles and push the actors they are addressing to believe them. During this process, scammers adopt attitudes, behaviors and mannerisms to make the scam scene more real, serious and natural, and to appear as normal and natural as possible. In the imagination of the victims, this staging is a sincere and honest reality, while in reality it is deceptive and fraudulent. Obviously, in the personalization of the scam scene, there is a desire to "appear", to pretend to be someone else. It is therefore a kind of staging orchestrated by the mobile crooks where each one plays a fraudulent role consciously in order to arouse positive feelings in the users and thus obtain money, the ultimate object of their investment.

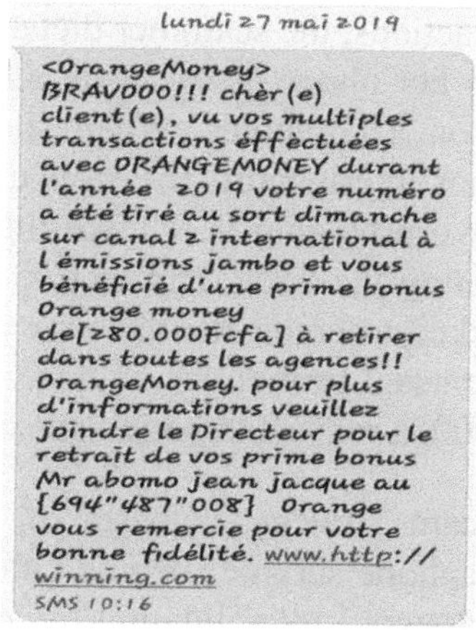

Figure 5.1. *Screenshot of a French money-making scam message. The message claims that the customer's name has been selected to win prize money*

5.2.2. *Enticing but misleading proposals*

This technique is based on the illusion of winning or the existence of important opportunities for the user. The scam is initiated by sending short message services (SMS) to cell phone users or by telephone calls to let them know that they have won alleged prizes in a draw or for their loyalty to their telephone network. Figure 5.1 shows an example of a message purporting to be from a mobile phone operator, which is similar to the messages people receive in Cameroon. Generally, these messages are similar to those delivered by the telecom companies, with the differences being the issue number, the contact number and the website. These elements, as reported by this user and many others, allow vigilant and informed people to avoid being taken in by the scam.

Some users may be contacted by phone call directly or as a result of a message. For example, one informant reported an attempted scam in which an unknown person, using an ordinary number, contacted her pretending to be the customer service department of an operator. After receiving a message about a lottery win, which she did not follow up on, she received a phone call a few hours later from a gentleman who was indignant that she had not called back the number indicated in the message to find out how to redeem her prize. Furthermore, after inquiring about the nature of their client's or target's phone and obtaining from her another number to call her back to facilitate the transactions, the scammer dictated a code to her that she was required to enter into her phone as she went along, making her actions appear to be controlled on a computer. When she became aware of the suspicious operation of her interlocutor, she stopped, which led to abusive exchanges between the two protagonists. Unlike this informant who, if she had cooperated by validating the code, had to pay the funds involuntarily, another informant was asked to credit her account with 100,000 francs that she had to transfer for the usual formalities, in order to enter into possession of a prize fund of 2,000,000, a request to which she did not respond.

Another telephone scam strategy consists of contacting users for a supposed update of a telephone service such as Orange money. The user is invited to change their password by validating the code #150*44#, in order to ensure the security of their account, after which they are invited to communicate the four-digit code received by message in order to complete the operation. This code, which is actually the payment code for online items, is then used by the mobile scammer to make purchases on the operator's partner commercial platforms.

The cell phone scams create tempting illusions by reappropriating the marketing strategies of telecommunication companies. More concretely, it relies on the "double vulnerability of the socio-technical chain of vigilance" elucidated in a study on the remote manipulation of Internet users by spam (Auray 2012, p. 112). The mobile scammers therefore seize the opportunity offered to them by the telephone to become part of the routine activities or patterns of users, through the counterfeiting of "normal" messages, that is, those from operators, which are regularly received by mobile phone users. Such a ruse relies on the regression of caution in the face of an ordinary activity. This mechanism is facilitated by the expansion of mobile money in Cameroon, which gives some users the opportunity to have mobile bank accounts that are similar to electronic purses and that they can de facto manipulate at will.

5.2.3. *Disguised telephone number confusion*

The phone number mix-up consists of sending a text message to the mobile phone operator and then letting them know that a sum of money has been mistakenly transferred to their mobile bank account. One of our respondents, who was a victim, reports the following:

> I received a message which said that I had received a deposit of 15,500 francs from a number. Well, looking online on the phone, I did not see the depositor. The message only appeared there with my name, and I did not open it. As I do business online, I thought surely, it's a customer who made the deposit and he will call me in the meantime. A few minutes later, a number called me and told me that a deposit had been made into my account. He really begged me to send it back, and even leave the withdrawal fee, and I said OK, I'll do it. Also, I said, I'm used to this kind of situation, because someone else made a deposit to my account by mistake and I sent it back. So automatically I sent the money back.

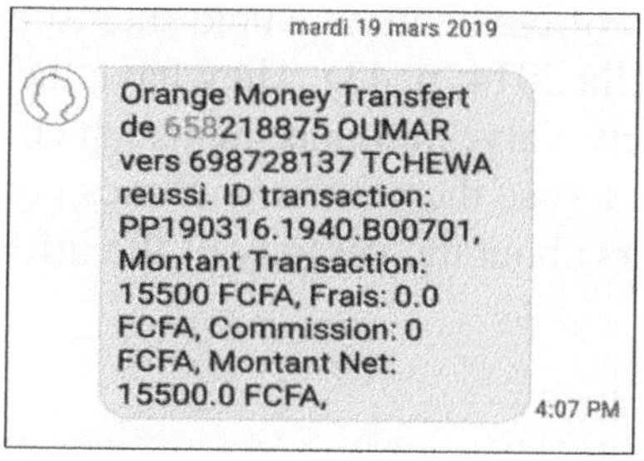

Figure 5.2. *Screenshot of a fraudulent money transaction message in French*

This story highlights the fact that mobile money, far from praiseworthy speeches, is at the center of the mobile scam. The mobile scammer holds the victim's name to make it easier for them to join the scam if they do not realize that it is an ordinary number that is sending the message. The mobile scam relies on the ordinary socio-professional concerns of the users to try to extort money from them. The mobile scammers use familiarization processes and language and discourse techniques that resort to persuasion mechanisms. They use a series of deceptive actions, based on sycophantic language, designed to induce the user to enter into an unfavorable commitment or to believe in the realities presented. To achieve their aims, they place their actions under the sign of charity, which "... is one of the universally shared values. Given its axiological, religious and social function, anyone who practices it is worthy of blind trust" (Hatolong Boho and Nkouda Sogpu 2014). The playing field for these actors is being expanded because of the freedom from geographic barriers and state borders brought about by the gift of the ubiquity of mobile telephony. It is a kind of transposition of ordinary deviance to a new fertile ground, namely technological tools and more precisely the cell phone. There is then a "co-evolution of technology and delinquency" (Dupont 2013). This de-sectorization of the spaces of deviance is indicative of the logic of conflict that information and communication technologies in general introduce into society, by means of the possible uses that they reveal. The nature of ICTs is not "crisogenic" and even less criminogenic, but their characteristics make them a real paradise of deviance. We thus agree with the idea of Karamoko (2015) that there is a link between cybercrime and the characteristics of the digital society, which are mobility, portability, the zapping society and techno-power.

Aware of the deviant nature[6] of their activity, mobile harassers, therefore, devise well-thought-out strategies to protect themselves or "hide what they are doing from the eyes of outsiders to shield themselves from any negative social sanction" (Ella-Ella 2014, p. 43). They have the skills that allow them to create a certain "safety valve" around them, which is necessary to safely commit their acts. Such a ruse therefore sometimes creates confusion among some mobile phone users about the veracity of the information received.

6 It should be noted that the phenomenon of "cell phone scams" comes under cybercrime, as defined by Law No. 2010/012 of December 21, 2010 on cyber security and cybercrime in Cameroon. From this point of view, it is a risky activity.

The risk of mobile phone scamming thrives in Cameroon in a context where there are structures and provisions for the protection of consumers who are cell phone users. It is therefore necessary to make an overview of the socio-institutional actions related to the security of mobile phone users.

5.3. The dynamics of cell phone use in Cameroon

In a risk situation, there is a target (mobile phone users), the risk bearers (mobile phone scammers) and the controllers or supervisors (the management structures). This section therefore focuses on the controllers and more specifically on the mediation practices at work not only to raise awareness but also to act in the face of the risk of mobile stalking.

5.3.1. *The Ministry of Posts and Telecommunications*

This is the government body responsible for the development and implementation of Cameroon's ICT policy. It ensures, among other things, the development of ICT and electronic communications in collaboration with other administrations, including operators and regulatory agencies of telecommunications and ICT. An interview with the Director of Network and Information Systems Security at MINPOSTEL revealed that the organization's action is basically at the legal and institutional levels. As such, three laws aimed at regulating, controlling and sanctioning cybercriminal acts were adopted in 2010.

Similarly, ANTIC and ART participate in securing cyberspace and the regulation of ICT and telecommunications. This Ministry examines the difficulties faced by regulators with a view to proposing technical solutions through the development of texts and laws and monitoring their implementation by regulators. However, the above-mentioned person in charge underlines some operational actions that aim at calling on users to be more careful, such as public awareness campaigns, public and media communications and the radio program "PPT performance". Similarly, the introduction of the identification of telephone subscribers by operators is part of the supervision of the cell phone sector in Cameroon.

5.3.2. *Agence Nationale des Technologies de l'Information et de la Communication*

ANTIC is the Cameroonian structure that guarantees the security of cyberspace in Cameroon. Through the CIRT, it ensures the security watch on the national cyberspace in synergy with other States. On the one hand, it has a preventive mission by raising awareness among ICT users on the risks related to various uses. To do this, it issues bulletins and security alerts on its website, organizes the annual National Forum for Internet Governance and educates the population through the radio program "Antic.cm", broadcast on the national radio. ANTIC also has a curative mission, because, as a CIRT staff member points out:

> Any criminal act that has been committed using ICT, any person who has been a victim of this kind of act can come and file a complaint at ANTIC, and we will assist in the investigations to find the perpetrators of these crimes.

This assistance to users is provided in collaboration with the security forces and cell phone operators. However, the staff recognize that the visibility of the actions of this structure remains insufficient in view of the growth in information and communication technologies in Cameroon. This organization also has two toll-free numbers, 8202 and 8206, to allow ICT users to report the abuses they experience on a daily basis.

5.3.3. *Agence de Régulation des Télécommunications*

ART is one of the two regulatory structures of telecommunications in Cameroon. It is, like ANTIC, placed under the supervision of MINPOSTEL. In the ICT sector, it is responsible, among other things, for ensuring the application of legislative and regulatory texts, sanctioning the failure of operators to meet their obligations and ensuring consumer protection. On this last point, the Director of this agency, in a statement in 2016, advised phone users to beware of scams. This agency can be likened to the gendarme of cell phone operators, because it ensures compliance with the standards generally prescribed by international bodies and by the specifications of operators as well as the requirements in force in terms of consumer health and safety. It thus supervises respect of the regulations concerning the identification of telephone subscribers. When an operator fails to comply with the standards set, ANTIC imposes financial penalties, such as those imposed on MTN, Orange and Nexttel in July 2017 and 2019. They were accused of selling pre-activated SIM cards and failing to comply with the number of SIM cards per subscriber and the standard relating to the identification of subscribers.

5.3.4. *Cell phone operators*

The mediation of cell phone operators, notably MTN and Orange, is both preventive and remedial in nature. In terms of prevention, they carry out awareness campaigns and educate users on what to do if they receive recommendations from anyone, either by call or by message, in order to reassure themselves of their veracity. Cell phone users are asked to avoid giving out their mobile money account codes, reacting on the spur of the moment by paying money to strangers and validating codes without being assured of the purpose and sender of any call or message. Telecommunication companies communicate via SMS alerts to their customers, through their websites and pages on digital social networks as well as through other media. For example, one of the messages sent to subscribers of MTN's mobile money service is the following:

> Beware of scams! When you receive a message inviting you to validate a transaction, read the content carefully before taking any action. Be vigilant!

Similarly, an awareness message from Orange has been broadcast on television since September 2019 to raise awareness among their customers.

When a subscriber is duped, these operators act as soon as they are informed, either to repair the damage immediately if it is still possible or to track down the scammers with the help of the security services. One of the Orange customer service agents tells us that in the case of a scam or an attempted scam by message, subscribers are asked to take a screenshot and send it by private message via the Facebook page or the WhatsApp number of the company, or call customer service to report what has happened.

It should be noted that the representatives of the operators in the field, commonly called "call boxers", help some users avoid situations of fraud, especially when they are contacted for information on dubious proposals or for money deposit operations.

> The first sign I saw was that as soon as he arrived, he first made a deposit of 35,000 in such a rush, and I said hey, what's the rush! A few minutes later he comes running back with 1000000, and I asked him, why the race? He said just send it to me and I'll tell you after. As it is my brother, I took the money and pretended to send it. After he let me know that his son was in trouble at the port and was stuck at customs, and I suspected that this was a scam since the number was withheld… In the end, the client, who is a relative of mine, realized the scam he'd been subject to and saw his losses reduced, as I had withheld his second payment.

This account by a communication credit saleswoman highlights her role in protecting certain clients. She also points out that, when faced with a client who is rushing to conduct a transaction, she urges them to be cautious about the scam situations that are out there.

5.3.5. *The judicial system and cell phone scams*

In Cameroon, Law No. 2010/012 of December 21, 2010 on cybersecurity and cybercrime (Law 2010)[7] acts as a theoretical tool to fight against digital risks. This legal aspect is important in understanding the institutional dynamics of mediation in favor of digital risks. This law aims to establish trust in electronic communication networks and information systems, the establishment of the legal regime of digital evidence of security activities and the protection of fundamental rights of individuals, including the right to human dignity, honor and privacy, and the legitimate interests of legal persons. The electronic communication networks covered by this law include, among others, terrestrial networks and electronic networks when they are used for the routing of electronic communications. Mobile crime as a digital risk is therefore part of cybercrime.

7 Law No. 2010/012 of December 21, 2010 on cybersecurity and cybercrime, available at: http://www.art.cm/sites/default/files/documents/loi_2010-012_cybersecurite_cybercriminalite.pdf [Accessed 15 September 2019].

At an operational level, the Cameroonian Gendarmerie and Police are state institutional structures, judicial in nature and designed to fight cybercrime in general and mobile scamming in particular. They are part of the regulatory mechanisms designed to identify possible deviations related to the use of mobile telephony in order to rectify the damage suffered by users. In this mediation task, they work in collaboration with cell phone companies in Cameroon (MTN, Orange, NEXTEL, etc.) and ANTIC. Their actions focus on recording complaints, investigations and arrests for submission to the criminal justice system. In the case of complaints relating to mobile scams, the procedure consists of requisitioning the mobile telephone companies in order to obtain certain information on the respondent. In particular, they aim to obtain from these companies the acts of identification, location and tracing of any phone number or phones that have been used in the process. The listings of incoming and outgoing calls and SMS are used to shed light on the chain of participants in the process and the numbers of people with whom the mobile scammer communicates the most in order to catch him. This mediation benefits from the support of ANTIC when the cell phone companies are late in responding to the requisitions made by the Judicial Police Officers. In this mediation operation, as a police officer in Yaoundé tells us, sometimes means other than traditional judicial and investigative techniques – notably digital social networks – are used to ambush the mobile scammers. Such a strategy corroborates the findings of Dekane (2016), which shows that information and communication technologies are helping to reconfigure both urban crime and the mechanisms for securing cities.

5.3.6. *Cell phone users and consumer associations*

Peer groups are important for some users as they help to convince cell phone operators of the scams they are being subjected to on a daily basis. Some of the people we interviewed said they were "saved" by people they knew and had confided in. The members of an interpersonal network, based on their experiences, help to prevent cell phone scams. Moreover, awareness messages are transferred by phone users to Internet platforms such as groups, Facebook pages and WhatsApp. However, civil society actions specifically focused on protecting cell phone users from the risk of scams are still non-existent. The associations listed by ANTIC are more interested in the risks of Internet scams. On the other hand, the Cameroonian League of Consumers, which is an association designed to protect consumer rights, sporadically moves to denounce the quality of the telephone network and the cost of calls.

It is thus established that there is a certain socio-institutional mobilization that intends to ensure the security of cell phone use. In view of the recurrence of the mobile phone scam despite these moderation initiatives, it is important to question the mode of governance of this risk.

5.4. Socio-institutional governance of cell phone use in Cameroon: Optimal or approximate mediations?

Lawrence E. Cohen and Marcus Felson showed in 1979 that for a crime to be committed, three elements are inescapable: a motivated offender, an attractive target and the absence of effective guards or supervisors. Without taking the proposal of these authors at face value before any analysis, we question in this section the forms of mediation at work to combat cell phone scams. This section shows that the socio-institutional governance of cell phone use faces various constraints that favor the prevalence of cell phone scamming.

5.4.1. *Information deficit of the users*

In view of the risk of cell phone scams that prevails in the mobile telephony sector, several actions are being implemented, as described above. However, there is a weakness in awareness-raising practices, not only in terms of awareness-raising channels but also and above all in terms of the frequency of awareness-raising activities. The components of the telephone use control system generally mobilize the media, with an average of only three types of media to choose from. The messages disseminated via websites, users' phones, digital social networks, radio, television and press releases are generally at the center of prevention, depending on each organization. Operational actions where these different institutions go to meet cell phone users are almost non-existent. The operators and their commercial representatives, aware of the importance of having a direct relationship with the client, regularly go out into the field to sell the SIM cards and popularize the offers, thus serving their own economic interests. However, these technological awareness campaigns cannot reach all categories of users because of cultural barriers, level of education and difficulties of access to and/or use of various media.

Another shortcoming is the frequency of awareness-raising activities, which are carried out on an ad hoc basis, in a context where mobile scammers are constantly updating their modus operandi. The focus is more on Internet-related risks, as far as the actions of structures such as ANTIC, ART and MINPOSTEL are concerned. Consumer associations, on the other hand, reduce their actions to demands for a better quality of telephone network and a reduction in communication costs without focusing on users' level of competence.

Moreover, the data collected highlights an information deficit on what to do in case of (attempted) scams. Indeed, cell phone users do not always contact the appropriate place or follow up on their complaints with the relevant departments. This situation reduces the ability of regulators and judicial services to act.

5.4.2. *Insufficient means of action*

Mediation practices require a certain number of resources, notably financial, technological and human. On the financial level, interviews with police and gendarmerie officers reveal the difficulties of implementing activities due to budgetary constraints. They emphasize, for example, that searches and raids in the field are costly and also require appropriate logistical means that these structures do not always have at their disposal. In addition, the lack of technological tools and the problem of retraining judicial personnel reduce the response capacity of regulators and judicial structures. Technological evolution, a catalyst for digital deviance, inevitably requires the updating of judicial investigation techniques, in an environment where state institutions for the protection of citizens and their right to access and use ICTs are not sufficiently equipped to deal with these new deviations.

It is important to emphasize that the impossibility of carrying out repressive actions in a solitary manner reduces the response capacities of various mediators. The investigations carried out highlight problems in coordinating response activities to recorded cases of cell phone scams. For example, ANTIC sometimes calls on the police to carry out searches and they do not react in time. As a result, the victim's case is abandoned due to the lack of immediate response.

5.4.3. *Mis-selling of SIM cards by mobile operators: An "ingredient" of mobile scammers*

The issue of subscriber identification is at the heart of recriminations by MINPOSTEL and ART. The mobile operators are generally accused of not controlling the process and of focusing on their economic objectives. These cell phone companies are, in a way, indirectly complicit in mobile scams in Cameroon. While the phone numbers are identified, nothing is done to find out if the person who identifies himself with a phone number is the actual user of that device. Sometimes it is SIM cards whose numbers are fraudulently pre-identified that are marketed. This imbroglio is advantageous for mobile scammers and allows them to hide their acts.

The obvious risk equates to a crime. Thus, in a context where socio-institutional organizations mobilize to cancel the opportunity offered by the telephone to mobile harassers, we assert, drawing on the approach of opportunities and routine activities (Cohen and Felson 1979), that the act requires a set of skills necessary for concealment. To do this, mobsters tend to use untraceable numbers. The SIMs are obtained in the streets of various cities in Cameroon at low prices, or even free of charge, provided that the applicant adds a certain amount of credit. The usual identification procedures are not respected, as these scammers use SIM cards that are not identified with their name. After a successful scam operation, the SIM cards used to perpetrate the act are destroyed to prevent them from being traced. The major cell phone user identification campaign launched by the Cameroonian government in 2016 aimed to assign to each number a user and an identity.

Mobile scammers are imaginative. The information they provide to identify these SIM cards comes from collected and stolen ID cards. Some of them are computer experts who specialize in making fake national ID cards used to create accounts. Many others contact street vendors who divert the vigilance of some buyers by using their ID cards to identify more than one SIM without their knowledge. The numbers identified in this way are used to contact the victims and to create mobile money accounts to receive the proceeds of the scam. Others use withheld numbers that do not allow other subscribers to contact them again. Mobile scam operations are based on well-structured teamwork where everyone plays their role in order to achieve the desired objectives. In addition, some are specialized in withdrawing the fruits of their labors. They are independent workers who are paid on commission. They intervene at the very end of the operations when it is necessary to withdraw the money from the scam.

5.4.4. *The ease of monetary transactions*

Mobile money transfer is characterized by the speed of operations due to the absence of any prior procedure, especially with regard to filling out forms, the need to have a national identity card and to have the password received from the initiator, as is very often required in traditional money transfer agencies. While the operators of this service have long emphasized this speed as an essential asset, it goes without saying that this easing facilitates the commission of acts by scammers. Moreover, the security measures taken by government institutions and telephone companies to identify telephone numbers and update mobile money accounts are often circumvented by some cell phone operators. Although these measures are much vaunted by those who introduce them and are intended to give each telephone number or each account an "identity", they are not always effective. Sometimes, a phone number is registered to one person and the account that is linked to it is registered to another. In view of this, it is clear that in terms of their content, these security measures do not really make anonymity disappear and are in fact a catalyst for mobile scams. In mobile money agencies, you generally do not need to fill out anything or answer any questions to receive your money. All you have to do is give your phone number and validate your code. These methods of withdrawing money are simple, easy and offer the mobile scammers the advantage of freedom, which is a bargain to them in terms of the leeway they have. The technological flaws of the mobile money system are therefore used by mobile hackers to carry out their acts.

5.4.5. *Technological constraints and border porosity*

> There are things that technology allows you to do no matter what you do, no matter what means you have.

These words from a CIRT employee highlight the fact that technology has loopholes that scammers seize upon to perpetrate their acts. Furthermore, the porous nature of borders makes it difficult to fight the phenomenon of mobile scams. An act committed from a foreign country requires deployment beyond borders, which has proven difficult up until now and requires collaboration with foreign services as well as enormous financial means for the searches.

These various limitations that emerge from the mediation actions are at the root of the failures observed in the socio-institutional governance of cell phone use in Cameroon.

5.5. Conclusion

This chapter addresses the issue of the socio-institutional governance of cell phone use in Cameroon in a context of the emergence of scams. It shows that this technological tool, far from the revolutionary character that has long been associated with it, is at the origin of a risk that we call mobile scamming. This research brings to light the socio-digital skills of mobile hackers who develop mechanisms inspired by the mastery of the socio-economic context of the environment in which their "clients" evolve, the marketing strategies of telecommunication companies and the possibilities of using the technology for other hostile purposes. Despite the existence of various socio-institutional mediations that carry out both preventive and remedial actions, such as MINPOSTEL, the judicial services, regulatory bodies, telecommunication companies and other users, the risk of mobile deception remains prevalent. This state of affairs has raised questions about their effectiveness in monitoring or supervising the mobile telephony sector and more specifically the mobile money service. Ultimately, this study reveals the pitfalls faced by socio-institutional organizations in their surveillance efforts, including technological constraints, insufficient means of action and the ever-imaginative strategies of fraudsters. These difficulties are not conducive to optimal mediation with respect to the risk of mobile scamming. However, beyond the approach of opportunities and routine activities according to which one of the reasons for the arrival of a crime lies in the absence or inefficiency of supervisors (Cohen and Felson 1979), this work takes on the unpredictable aspects of certain crimes, because whatever the level of security of the telecommunication system, zero risk does not exist. The technological and environmental constraints favor the commission of certain acts of mobile crime. Thus, the latter results from the opportunity offered by the influence of mobile financial transactions in the daily life of a marginal group of cell phone users, from the socio-institutional surveillance that is subject to various constraints, as well as from the motivation and skills of the mobile crooks to seize this opportunity. This study therefore reveals the challenges to be met in order to optimize the supervision of the mobile money sector in Cameroon, among which raising awareness among cell phone users to help them discern the information that reaches them by cell phone on a daily basis appears to be crucial in the fight against this type of cell phone scam.

5.6. References

Abia, W.A., Jato, D.M., Agejo, P.A., Abia, E.A., Njuacha, G.E., Amana, D.A., Akebe, L.K., Takang, A.S.J., Ekuri, D.O. (2010). Cameroonian youths, their attractions to scamming and strategies to divert attention. *International NGO Journal*, 5(5), 110–116.

ART (2018). Observatoire annuel 2017 du marché des communications électroniques, Cameroun. Agence de Régulation des télécommunications, Yaoundé.

Auray, N. (2012). Manipulation à distance et fascination curieuse. *Réseaux*, (171), 103–132.

Bauer, R. (1960). Consumer behavior as risk taking. In *Dynamic Marketing for a Changing World*, Hancock R. (ed.). Marketing Classics Press, Decatur, GA.

Beck, U. (2008). *La société du risque : sur la voie d'une autre modernité*. Flammarion, Paris.

Benbouzid, B. and Peaucellier, S. (2016). L'escroquerie sur Internet : la plainte et la prise de parole publique des victimes. *Réseaux*, (197–198), 137–17.

Bogui, J.-J. (2009). Usages et appropriation des TIC par les jeunes ivoiriens : de l'espoir au désenchantement. *TIC & Développement*, 4(2008–2009).

Bogui, J.-J. and Atchoua, N.J. (2016). La régulation des usages des TIC en Côte d'Ivoire : entre identification et craintes de profilage des populations. *TIC & société*, 10(1), 1–17.

de Certeau M. (1990). *L'invention du quotidien, tome 1 : arts de faire*. UGE Éditions, Paris.

Cheneau-loquay, A. (2008). Rôle joué par l'économie informelle dans l'appropriation des TIC en milieu urbain en Afrique de l'Ouest. *Netcom*, 22(1/2), 109–126.

Chevanaz, R. and Paraschiv, C. (2011). Processus de rencontre sur Internet : une étude empirique de la perception du risque. *Management & Avenir*, 44(4), 124–146.

CIRT (2018). Welcome to CIRT [Online]. Available at: https://www.cirt.cm/ [Accessed 2 January 2020].

Cohen, E.L. and Felson, M. (1979). Social change and crime rate trends: A routine activity approach. *American Sociological Review*, 44(4), 588–608.

Cross, C. (2018). Victims' motivations for reporting to the "fraud justice network". *Police Practice and Research*, 19(6), 550–564.

Dekane, E (2016). Le banditisme urbain à l'ère des TIC : un appel à la prudence et aux nouvelles méthodes de sécurisation des villes camerounaises. *Revue sciences, langage et communication*, 1(3), 1–18.

Dupont, B. (2013). La coévolution de la technologie et de la délinquance : quelques intuitions criminologique. *Annales Internationales de Criminologie*, 51(1/2), 39–56.

Ella-Ella, S.B. (2014). *Quand le capitalisme cynégétique envahit la réserve du Dja, étude de la sociologie de la chasse déviante*. PUY Éditions, Yaoundé.

Goffman, E. (1974). *Les rites d'interaction*. Minuit Éditions, Paris.

Hatolong Boho, Z. and Nkouda Sopgu, R.V. (2014). Stratégies argumentatives du genre cyber-épistolaire : rhétorique de l'arnaque et (en)jeux de faces. *Multilinguales*, 2014(4), 50–75.

Karamoko, T. (2015). La société digitale et les racines de la cybercriminalité. *Perspective Philosophique*, (009), 1–19.

Kermisch, C. (2012). Vers une définition multidimensionnelle du risque. *VertigO – la revue électronique en sciences de l'environnement*, 12(2) [Online]. Available at: https://doi.org/10.4000/vertigo.12214 [Accessed 2 January 2021].

Loukou, A.F. (2012). Les TIC au service du développement en Afrique [Online]. *TIC & société*, 5(2–3) [Online]. Available at: https://doi.org/10.4000/ticetsociete.1047 [Accessed 2 January 2021].

Nji, B.E. and Ebi, N.J. (2019). Cyber criminality and electronic banking in Cameroon: What prospects? *National Journal of Cyber Security Law*, 2(1), 23–37.

Nkongho, A.M. (2019). Cyber insecurity in e-commerce transactions in Cameroon: What prospects for the digital economy. *National Journal of Cyber Security Law*, 1(2), 16–32.

Perrot, T. (2015). Escroqueries et arnaques sur Internet au Ghana : le phénomène sakawa. *Les Enjeux de l'information et de la communication*, (15/2B), 43–50.

Rengou, A.B. (2017). L'arnaque cybernétique : analyse d'un phénomène en pleine expansion dans le Noun. Master's Sociology Thesis, University of Dschang, Cameroon.

Reyns, B.W. and Randa, R. (2015). Victim reporting behaviors following identity theft victimization results from the national crime victimization survey. *Crime & Delinquency*, 63(7), 814–838 [Online]. Available at: https://doi.org/10.1177/0011128715620428 [Accessed 2 January 2021].

Tasso Boni, F. (2014). La cybercriminalité au Bénin : une étude sociologique à partir des usages intelligents des technologies de l'information et de la communication. *Les Enjeux de l'information et de la communication*, (15/2B), 35–42.

PART 3

Digital Risks: Practices and Mediation

6

Towards a Normative Prescription of Information Practices on Digital Social Networks: A Study of Documentary Pedagogical Projects in Middle School

6.1. Introduction

In the French National Education system, an institutional preoccupation with media and information literacy emphasizes the need to train students in order to "enable them [...] to exercise their citizenship in an information and communication society[1]". In fact, information practices pose a challenge to the educational community, which attempts to get a grip on its training issues.

Based on the observation that young people's information practices are concentrated on digital social networks (DSNs) and include communicative, information-seeking and social activities (Aillerie 2008), we focused on a specific context of use: digital social networks. After studying our corpus, we found that pedagogical implementations on digital social networks were often an educational response to digital risk. Thus, we would like to set out

Chapter written by Adeline ENTRAYGUES.

1 Ministère de l'Éducation nationale, de la Jeunesse et des Sports – Direction générale de l'enseignement scolaire. (2021). Éducation aux médias et à l'information. *EduSol* [Online]. Available at: https://eduscol.education.fr/1531/education-aux-medias-et-l-information [Accessed January 3, 2021].

the argument that the notion of risk is a common thread in the pedagogical practices encountered and shapes the construction of an information culture.

After defining the risk that concerns us, we will propose a model of our research objects, namely, heterogeneous information practices, DSNs and information literacy. After presenting our empirical methodology, we will show two axes of results relating to the representations of the teacher librarians and the learners of the risks within the framework of information practices on the DSNs. First, we will situate the risks as perceived by teachers and learners, and then we will move on to the treatment of risk in learning and its reception by learners. Finally, we will argue that risk influences the formation of an information culture on DSNs.

6.2. Contextualization of risk

To begin with, we want to differentiate between danger and risk; with a more generic and even almost emotional entry, danger is a "situation where a person's (or a country's) safety or existence is threatened[2]". Risk represents a possible danger which is more or less predictable, inherent to a situation or an activity[3] and comes to constitute a context or an environment.

We will thus speak about the digital risk, in other words, risk through a particular medium. In the case that concerns us, this is the digital social network. Young people's information practices lead to different forms of risks. First, informational risks concern the relationship to information and its content, such as the evaluation of information or the manipulation of information and/or images (Serres 2012). Digital risks can be ethical, with issues related to the protection of personal data, digital identity or e-reputation and the respect of others in their private lives (Merzeau 2013; Cardon 2015). Finally, there are legal risks related to the violation of copyright and cybercrime.

6.3. Issues to consider

From these multiform risks, it is advisable to cross the objects and the framework of use engaged to understand the training issues at stake. We will

---

2 Definition of danger from the *Trésor de la langue française informatisé* (TLFi).

3 Definition of risk from the *Trésor de la langue française informatisé* (TLFi).

try to question the link between plural information practices in the school sphere and the private sphere, a differentiation that we will make explicit in our theoretical framework, and we will then try to grasp the link between representation, practices and formation of an information culture. Indeed, why is the risk inherent to information practices on the DSNs so dominant in the documentary pedagogy focused on the DSNs? And in what way does the understanding of digital risks on DSNs in pedagogical projects impact the formation of a multiscale and stratified information culture?

6.4. Research objects

As an initial research object, we mobilized the plural and generalizing notion of information practices *as the entirety of relationships to information,* be they informational, communicational, socializing or playful. For Aillerie, they "refer to the act of being informed" (Aillerie 2011, pp. 99–100) and "are essentially to do with the relationship to knowledge maintained by the individual". (Aillerie 2010, p. 190). Chaudiron and Ihadjadene (2011), for whom informational practices are "the way in which a set of devices, sources, tools and cognitive skills are effectively mobilized in the different situations of producing, researching and processing information", focus on the search for information and omit the playful and social aspect of information. We consider communicative, playful and social practices to be an integral part of information practices, particularly on the DSNs where the relationship to information creates social and communicative practices that go as far as an affiliation link between peers (Dauphin 2012).

Two types of practices, prescribed and informal, come into tension in apparently opposed contexts of use. Béguin-Verbrugge (2006) defines prescribed practices according to school expectations: "Practices prescribed by the school and modelled according to criteria of collective efficiency; informational output, but also cultural legitimacy".

Non-prescribed practices are constructed according to personal satisfaction: "Ordinary social practices, not prescribed or regulated by an authority, not explicitly structured, but effective in the satisfaction they provide on a daily basis".

There is a clear opposition between two fields of application, between an academic context and a personal context. Prescription concerns “an official framework of learning” (Aillerie 2010, p. 12) and echoes good practice in relation to forms of purpose recognized by the validating authoritative institution that is the school (Liquète 2012).

Informal or non-formal practices are expressed freely in contexts of personal use, but influence the school sphere (Soumagnac and Capelle 2017). We would like to return to this semantic distinction between informality and formality, which finds an initial theoretical basis in scientific research, and also offers insight into understanding the zones of influence and porosity of the private and school spheres. Some researchers, despite definitional dissensus, question the relevance of this differentiation (Le Deuff 2011; Aillerie 2016). Nevertheless, we have chosen to maintain this distinction because it gives meaning to the research due to the context of use, and it seems important to understand the stakes in the practices and representations on the DSNs, of the influence of prescriptive normativity in the information practices on the DSNs.

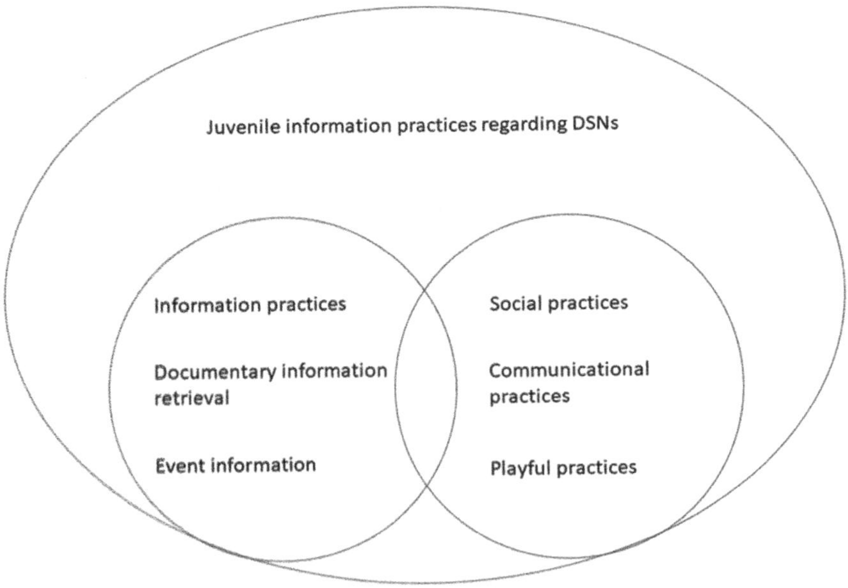

Figure 6.1. *Modeling information practices on DSNs*

Socio-numeric networks, our second mobilized object between flow, information and platform (Ellison and Boyd 2013), are at the heart of these juvenile information practices, directly related to the need for sociability inherent to the construction of the juvenile identity (Cordier 2015), and to self-exposition (Cardon 2008) to the point of modifying the cultural and communicational reference points (Pasquier 2005).

> A social network site is a networked communication platform in which participants 1) have uniquely identifiable profiles that consist of user-supplied content, content provided by other users, and/or system-provided data; 2) can publicly articulate connections that can be viewed and traversed by others; and 3) can consume, produce, and/or interact with streams of user-generated content provided by their connections on the site… (Ellison and Boyd 2013).

Stenger and Coutant (2011) define social media as "a group of online applications that are based on the ideology and technology of Web 2.0 and allow the creation and exchange of user-generated content". The mediation of social interaction gives digital social networks their specificity. Mediated sociability is an undeniable part of what DSNs call "attractiveness essentially (based) on the opportunity to find and interact with one's friends through profiles, contact lists and applications across a wide variety of activities" (Ibid., 13).

The plural notion of information culture that we pose here as our third object of research imposes itself as a thread in the framework of our reflection. Our theoretical framework questions, above all, the form of *information culture*. In order to better understand the intrinsic stakes, we pose a dual model with two complementary currents; information culture appears, on the one hand, as a common and general culture (Doueihi 2011), and also constitutes an undefined form of school culture, on the other hand. Conceptual understandings such as information culture, informational culture, information mastery or information literacy seem to be torn between social practices and theoretical notions (Le Deuff 2012). The task proves complex due to the lack of consensus around a polysemous (Chante 2010) or even catch-all concept (Serres 2010) in perpetual instability (Le Deuff 2010).

However, we wonder if a unique re-definition of information culture remains indispensable in the sense that it brings together two complementary, but no less different, visions of social and pedagogical practices, which amalgamate education and society, skills and knowledge. It thus mixes social, political and educational issues and aligns with the inescapable worries that people have in a changing information environment:

> Media literacy and ethical consideration [...] implies social integration (Juanals 2003, pp. 24–25).

6.5. Research protocol

Our empirical methodological choices are based on qualitative research as part of a comprehensive approach, in order to recompose the intentionalities of the subjects. This comprehensive and systemic approach (Morin 1990; Watier 2002), based on a sociological reflection of social practices, allows for an all-embracing approach at the intersection between interpretation and explanation.

Our discursive approach crossed with non-participant observation of pedagogical projects in the classroom is based on a corpus of multiple documents, including transcripts of comprehensive interviews of learners and teachers, observations of pedagogical sessions, institutional and professional documents and pedagogical documents collected during our observations: it thus allows for the comparative cross-referencing of data from this methodology. To complete our approach to discourse, we used the textual analysis software, Tropes[4], to deepen the semantic work on the arrangement of discourse.

Our field of observation is eight middle schools spread over the national territory[5], including a pedagogical project carried out with a teacher librarian on a DSN; our corpus includes two target audiences: the teachers (teacher librarians) and the learners (the students). As far as the teachers are concerned, there are 11 teacher librarians employed at the school where the project takes place. The classes are of different grade levels, from sixth to tenth grade. We interviewed about 10 students per class, which constitutes a corpus of 81 students. We chose to create unique interview grids for all of the students from sixth to tenth grade, in order to respect a scientific

4 Communauté Tropes (2013). Tropes version 8.4 : manuel de référence. *Tropes* [Online]. Available at: https://www.tropes.fr/ManuelDeTropesV840.pdf.

5 Our observation sites are spread over five academies.

methodology; however, our comprehensive and semi-directive approach allowed us to adapt to each student, particularly during the more theoretical questioning. The interviews of the learners were designed to focus on their personal information practices, their perception of the DSNs and the information society, and finally, the relationship between DSNs and schools, and their expectations in terms of learning. For the teacher librarians, we organized our interview grid around three axes: first, information culture and the role of the DSNs; second, the objectives of the prescribed practices implemented with the informal juvenile practices and DSNs; and third, a description and assessment of the main pedagogical project on the DSN. Observations were reflexively fed back into interviews for learners and teachers. We attended six pedagogical sessions in the CDI, maintaining an external position, and proceeded to record the whole session, for which we used the most significant moments.

	Type of establishment	**Geographic location**	**Students' age**	**Project**	**Number of interviews with students**	**Session observation**
ES1	High school	Alsace	11–12	*Twictée entre deux classes de sixième*	11	YES
ES2	High school	Nouvelle Aquitaine	12–14	*Sensibilisation sur l'estime de soi et l'identité numérique* (7th grade) *Sensibilisation sur la rumeur et le harcèlement* (8th grade)	9	YES
ES3	High school	Nouvelle Aquitaine	12–13	*Identité et présence numérique*	7	YES
ES4	Senior high school	Ile de France	15–16	*Réseaux sociaux et identité numérique*	4	NO
ES5	High school	Ile de France	14–15	*Création d'une grainothèque et communication sur Twitter*	7	YES
ES6	High school	Centre-Val de Loire	14–15	*Flash Tweet Edu*	12	YES
ES7	Senior high school	Auvergne-Rhône-Alpes	15–16	*Rencontre littéraire sur Twitter*	10	NO
ES8	High school	Nouvelle Aquitaine	11–15	*Défi-Babélio*	21	YES

Table 6.1. *Description of the eight observation sites*

Through this methodology, which allows a scientific approach, and also a differentiation relative to the specificities of our observation fields, we were able to understand the personal or school information practices and the representations in a context of formality and informality on the DSNs of learners and teachers.

6.6. Risk regarding DSNs in the pedagogical approach

The analysis of our corpus leads us to consider a documentary pedagogy guided by risk awareness, on the one hand, in terms of perceptions, and the implementation of pedagogical projects on the DSN, on the other hand.

6.6.1. *Raising awareness of risks: an obvious approach for teacher librarians*

The institutional recommendations and the professional missions of teacher librarians converge towards a mission to train citizens of the information society:

> By diversifying resources, methods and tools, [the specialist teacher librarian] contributes to the development of critical thinking with regard to sources of knowledge and information. They take into account the evolution of students' information practices and place their action within the framework of media and information literacy (MEN 2017).

For the pedagogues represented by the teacher librarians, the subject of DSN implies awareness-raising pedagogical sequences, even though the projects observed have also chosen to approach DSNs as media or communication vectors. Three of the eight projects observed (ES2; ES3; ES4) have adopted a theoretical approach based on risks, with awareness-raising on three forms of risk: those linked to social relations with rumors and harassment (ES2), and those linked to self-protection and digital identity (ES3 and ES4); moreover, the eight projects observed have opted for prescriptive objectives, even though the aims of the information practices engaged in on the DSN may be informational or communicational. It is a question of addressing the danger that the use of a DSN represents: in all of the discourses of the teacher librarians, we have noticed the use of the term risk and/or dangers to speak about the use of a DNS in the school context:

> We present the risks, all of them, and an awareness of the risks (ES4DOC).
>
> It's when we talk about social networks, the dangers, the risks, what not to do (ES6DOC).
>
> Let's say we still have to warn them about the risks that these DSNs bring and our whole role is more prevention, I think that's what's important (ES7DOC1).
>
> It allows you to work on all of the DNL codes I was telling you about, the profiles, the little intro, the presentation, making friends, seeing the comments while avoiding all the risks (ES8DOC).

However, some reluctance persists and some teacher librarians in ES4 and ES6 regret the lack of effectiveness of these often-stigmatizing methods of training: stigmatization attenuates the effect of these learning points and reduces the DSN to a dangerous space:

> And so, we totally leave out the positive effects for citizens. The students dig their heels in and get fed up because they don't see the point (ES4DOC).
>
> We have too much of a negative input, when actually, social networks are many other positive things (ES6DOC).

6.6.2. *Considering the views of learners and teachers*

To go further in our conclusions, we wanted to consider the terms used in the collected verbatims, which reveal the representations of the learners and teachers: we worked on the vocabulary related to risk and danger and tried to understand the reasons in terms of causes and consequences. With the Tropes software and the star graph, we analyzed the relationships between the semantic contents organized by the scenario, by indicating the position of the classes in relation to the central class, the types of relationships, upstream or downstream.

Below, we propose two translated star graphs from the Tropes software, which confirm that the prescribed or personal information practices are found challenging by the subjects.

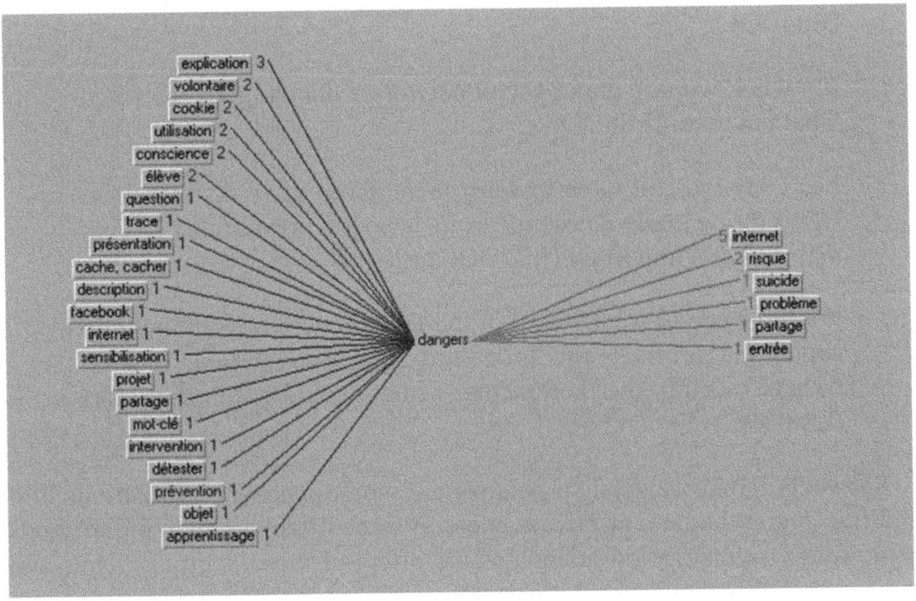

Figure 6.2. *Star graph of dangers*[6]*. For a color version of this figure, see www.iste.co.uk/capelle/digitalrisks.zip*

There are 48 cumulative occurrences relating to dangers and risks, which show that this aspect remains prevalent for the two audiences interviewed. Talking about DSNs for our two audiences means avoiding risks, or at least thinking about them. The meanings related to danger and risk highlight that education plays an essential role in the consideration of risks. We also point out that risks and hazards remain very generic, suggesting that subjects repeat established discourses.

6 Translation from left to right and top to bottom: *explication* = argument, *volontaire* = headstrong person, *cookie* = cookies, *utilisation* = use, *conscience* = consciousness, élève = student, question, trace, présentation = presentation, *cache, cacher* = hiding, description, facebook, internet, *sensibilisation* = awareness, *projet* = project, *partage* = sharing, *mot-clé* = key word, intervention, *détester* = hating, *prévention* = prevention, *objet* = object, *apprentissage* = learning, dangers, internet, *risque* = risk, suicide, *problème* = problem, *partage* = sharing, *entrée* = input.

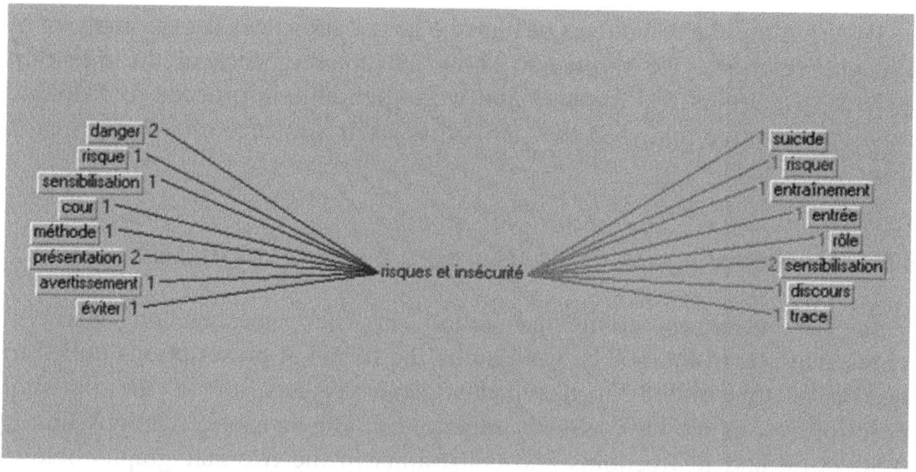

Figure 6.3. *Star graph of risks and risk factors*[7]. *For a color version of this figure, see www.iste.co.uk/capelle/digitalrisks.zip*

For learners, the risks mentioned often come from school and parental prescriptions common to a profession or a peer group, and seem inherent to the DSN. Emma (ES8EL15) told us:

> I pay attention to the people who follow me, I also look carefully at the people I follow [...] I also pay attention to what I publish, because I don't just publish any old thing.

For teacher librarians, dealing with the DSN is above all about raising awareness of the dangers and transmitting protection mechanisms. The ES4 teacher librarian thinks:

> It has to be repetitive to create automatic behaviors. That is to say, if a teacher in their school tells them that they have to be careful, it will not be enough if they are told that they have to be careful. Without showing them, it will be even less sufficient.

7 Translation from left to right and top to bottom: danger, *risque* = risk, *sensibilisation* = awareness, *cour* = schoolyard, *méthode* = method, *présentation* = presentation, *avertissement* = warning, *éviter* = avoidance, *risques et insécurité* = risks and insecurity, suicide, *risquer* = risking, *entraînement* = training, *rôle* = role, *sensibilisation* = awareness, *discours* = discourse, trace.

By grouping the discourses of our two target audiences for the analysis of semantic relations, we wanted to show the convergences of the intention systems of learners and teachers and were then able to proceed to a double cross-comparative analysis, that is, within each audience and then in each observation field.

6.6.3. *Considering the risks: learners aware of digital dangers*

In the verbatim comments found in the speeches on learners' representations of the DSNs, we assume the forms of prescriptions linked to the DSNs transmitted in the pedagogical projects, or in the parental prescriptions, as we have already mentioned. The semantic relations linked to attention and to freedoms and constraints in the two star graphs below, reveal blurred or even distorted representations of the DSNs and strong expectations in terms of education. The risks incurred on the DSNs result from the declared reasoned information practices, which, for the students means paying particular attention and respecting rules. To illustrate this, we present two star graphs related to attention and freedom and constraints.

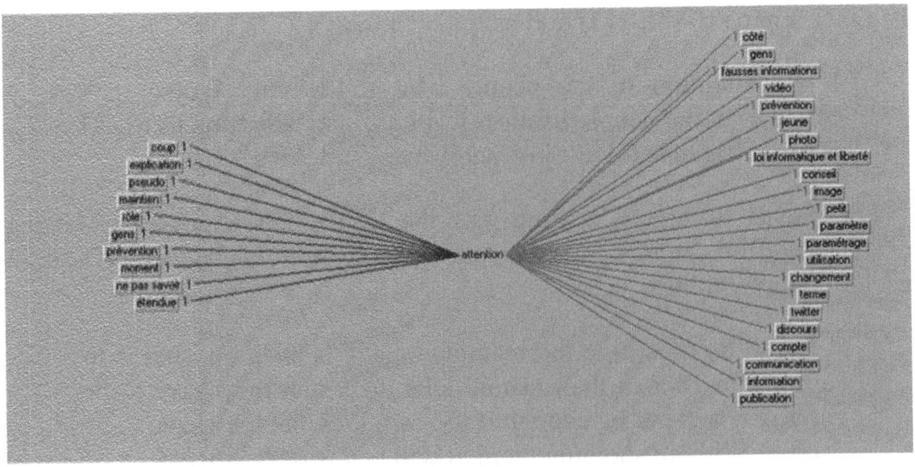

Figure 6.4. *Star graph of attention*[8]*. For a color version of this figure, see www.iste.co.uk/capelle/digitalrisks.zip*

8 Translation from left to right and top to bottom: *coup* = time, *explication* = explanation, *pseudo* = pseudo, *maintien* = maintenance, *rôle* = role, *gens* = people, *prévention* = prevention, *moment* = moment, *ne pas savoir* = not knowing, *étendue* = extent, attention, *attention* = warning, *côté* = direction, *gens* = people, *fausses informations* = fake news, *vidéo* = video,

Attention is related to people and publications. These three points regroup the network, content and photographs. We observe that both learning and the legal framework influence the relationship to attention; the image as a whole is a central concern.

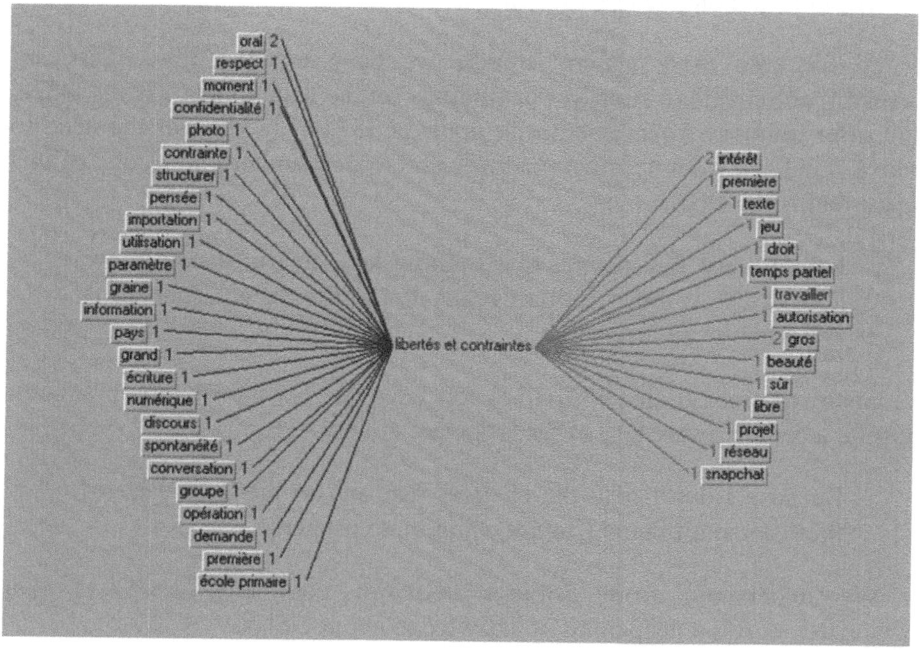

Figure 6.5. *Star graph of freedom and constraints*[9]*. For a color version of this figure, see www.iste.co.uk/capelle/digitalrisks.zip*

prévention = prevention, *jeune* = young, photo, *loi informatique et liberté* = computer law and freedom, *conseil* = advice, image, *petit* = small, *paramètre* = factor, *paramétrage* = configuration, *utilisation* = use, *changement* = change, *terme* = term, twitter, *discours* = discourse, *compte* = account, communication, information, publication.

9 Translation from left to right and top to bottom: oral, respect, moment, *confidentialité* = confidentiality, photo, *contrainte* = constraint, *structurer* = structure, *pensée* = thought, importation, *utilisation* = use, *paramètre* = factor, *graine* = potential, information, *pays* = country, *grand* = large, *écriture* = writing, *numérique* = digital, *discours* = discourse, *spontanéité* = spontaneity, conversation, *groupe* = group, *opération* = operation, *demande* = request, première = first, *école primaire* = elementary school, *liberté et contraintes* = freedom and constraints, *intérêt* = concern, *première* = first, *texte* = text, *jeu* = game, *droit* = law, *temps partiel* = part time, *travailler* = work, *autorisation* = authorization, *gros* = large, *beauté* = beauty, *sûr* = secure, *libre* = free, *projet* = project, *réseau* = network, snapchat.

On the other hand, the relations are reversed for the relations of the occurrences referring to the notion of freedom and constraints. Reasoned information practices on the DSNs imply a complex relationship to the standard and the rule for adolescents, and even more so when the prescriptions are dictated by the school.

To illustrate these general trends, we have noted a few significant verbatim comments to support our findings on the need to address the DSNs and offer training. Anaïs in seventh grade (ES3EL4) recognizes the need to take a step back from her own practices and to deepen her knowledge of this space that is the DSN:

> I think very few people really know how to use it; we know how to use it, but we don't really know the dangers behind it and what it can lead to.

Gaelle in ninth grade (ES6EL12) concedes that her information practices, despite a certain awareness of the risks, are not perfect:

> Because I don't read the charters that are on there, I've heard the risks, but I know I'm not using them properly.

We understand from Albin's verbatim comment (ES3EL1) that awareness allows a deepening of knowledge about DSNs:

> I knew a little bit about the suicide risks... but there was so much I didn't know.

Estéban (ES8EL17) clearly states that schools must address the DSN because it is a current problem:

> Yes, I think it's useful because now there are problems that arise in relation to DSNs, it's a current problem so I think we need to talk about it.

The expectations expressed by the learners in relation to their representations of the risks related to the DSNs remain, then, in agreement with the pedagogical approaches of the teachers.

6.7. Discovering DSNs in a school context: Dealing with risks

6.7.1. *Pedagogical projects on DSNs to prevent risks: Teachers' perspectives*

Setting up a pedagogical project on DSNs is often the result of a training requirement defined by the teacher librarians. The teacher librarian of the ES6 considers that it is necessary to train in the use of the DSN and provide an instruction manual to ensure its proper use:

> I use them as a learning object, the dangers the risks.

A teacher librarian in ES7 (ES7DOC2) shares the same view:

> This year, the school brought in a speaker to explain the dangers of the Internet.

The ES8 teacher librarian describes the same practices of digital media awareness:

> There is an awareness of the dangers of the Internet: the BPDJ (juvenile crime prevention brigade in France) comes and does a session with all of the 13–14 year olds and 11–12 year olds, which is not bad, but afterwards they really demonize the tool.

With the following two excerpts, we see that the pedagogical projects manifest themselves as a solution to a risk and aim to transmit prescribed practices into the private sphere. The teacher librarians of ES1 and ES7 set up the project on the DSN in response to a problem encountered in the school:

> If we don't train our students to respect good practices, good rules, we can quickly get into trouble – there have already been problems in our school (ES1DOC1).

> Allow students to use their cell phones in the schoolyard except when they leave the school. We realized that there were a lot of problems to deal with in terms of harassment, photos that are posted on the DSN, problems with students viewing videos (ES7DOC2).

6.7.2. *Overcoming risks: Learners' perspectives*

The trends we have observed point towards a modification of practices according to the sphere of use and prescriptions, and sometimes, a modification of perceptions, which leads to a change in information practices. Paula (ES8EL4) comments on the redefinition of her network:

> I learned to be more careful before I had a private account and I had a lot of subscribers, so I took them all off except my friends and people I knew.

Just like Manuel (ES1EL1) who seems to be more wary of "people who are behind another screen, because you never know who it might be. I'll be careful about that and I think that's the most important thing".

Anouk (ES6EL7), for her part, assures us that she is now more sensitive to the public/private boundary:

> Yes, a little bit because at the beginning we talked about privacy, that you shouldn't post too much, that you should be careful when setting the parameters of your accounts, so yes I'm going to be more careful about that.

Thomas (ES8EL18) recognizes the useful contribution of the project:

> Yes, it can help us learn hazard prevention.

Gaelle (ES6EL12) seems to be aware of the legal framework even though the fact that she is aware of it does not change her information practices:

> Because I don't read the charters that are on there, I've heard the risks, but I know I'm not using them properly.

In response to the prescriptions, Eloise (ES3EL6) reiterates what she has retained:

> We studied DSNs; we had to be careful with our image, in that we had to have the agreement of others to post photos and comments.

Imène (ES4EL1), despite a limited adhesion to the chosen DSN, declares that the school work on setting parameters for accounts has been really useful and concludes that:

> I don't like Facebook because I don't like the principle of Facebook. I said to myself I found it interesting, I said to myself everyone is on DSNs. We don't pay too much attention to what we post or what we don't post, but ultimately they can reuse our information to make a whole thing out of it. It's us who are in danger.

Adrien (ES5EL7) seems assured that he will change his practices according to the informational prescriptions learned during the project:

> Yes, I'll be more careful about what information people can share. Photos can be edited.

The students reappropriate the risks highlighted in the educational projects and adapt them between parental, school and personal prescriptions.

6.8. Perspectives for an information culture

The results presented above question the notion of information culture, which is inherent to an information and library didactic, especially on DSNs. Based on our initial reflection on the understanding of risks in pedagogical projects by teachers and learners, we propose some avenues for the modeling of an information culture to prevent risk.

6.8.1. *Risks, standards and education*

Educational approaches approach risk as a situation to be avoided with a critical and reflective approach; the school standard projects a coercive and prescriptive information culture. The risk situations experienced by the learners are divided into legal risk (what must be done), ethical risk (what should be done) and informational risk (what should be done to correctly satisfy one's informational need). The three dimensions of normative discourses, coercive, prescriptive and integrative, advanced by Capelle (2018) based on her analysis of teachers' perceptions, intersect with the pedagogical concerns identified in this study. Prescriptions can lead to the establishment of standardized practices with the aim of acquiring a common culture. Normativity guides the pedagogical approach to risk, and risk influences the need for reflexivity to foster learning, to the point of moving towards a normalization of risk. We argue that information culture is composed of prescribed academic norms and underlying social norms.

6.8.2. *A culture of information in training*

First, in view of our results, information culture is a culture under construction and emergent in the sense that it is formed separately from a social reality emanating from information practices and the perceptions of subjects. We draw attention to the fact that the practices observed are emerging practices, but are not representative of teaching practices; they do, however, have the merit of questioning the profession and stem from professionals who experiment with students. These timid attempts can be explained by the specificity of a moving technological object and its anchorage in a social reality that is complex to grasp: the difficulty of approaching DSNs in the school sphere favors the orientation of an information and library didactic through risk. From the point of view of the teacher librarians we met, the social practices of young people on DSNs constitute a need for support and prescription towards documentary mediation (Liquète 2010). Media and information literacy then naturally moves towards a prescriptive educational approach in relation to young people's information practices. For teacher librarians, pedagogical projects on the DSN respond to both training missions and societal necessities. For the learners, the pedagogical projects are adapted to their personal information practices.

Through our observations, we have noted a culture of information contextualized to a dichotomous knowledge between the private sphere and the school sphere, despite the will of the professionals to create a common culture transferable in both contexts of use. However, a form of common social awareness of risks and information practices influenced by the peer group and by the refusal of a normative prescription is emerging. Despite a desire to make globalizing information practices cohabit with an anchoring in social reference know-how, informational representations related to DSNs oppose information practices and consequently, informational learning (Serres 2017).

Information literacy is critical to support informal information practices and a need for risk distancing.

6.9. Conclusion

This information culture, which we describe as composite, responds to a need to accompany informal juvenile information practices, as much as it allows the acquisition of a critical mind. As we have seen, the intention systems of learners and teachers are convergent within this reflective approach. However, we question the paradox between a will to transmit a critical spirit and the prescriptive normativity of the forms of information practices observed. Hence, we propose the formation of a multiscale and stratified information culture.

6.10. References

Aillerie, K. (2008). *Les pratiques de recherche d'information informelles des jeunes sur internet. Communication*, Colloque international de l'ERTé, Lille [Online]. Available at: http://archivesic.ccsd.cnrs.fr/sic_00344181/ [Accessed 3 January 2021].

Aillerie, K. (2010). Les pratiques de recherche d'information informelles des jeunes sur internet. In *L'éducation à la culture informationnelle*, Chapron, F. and Delamotte, E. (eds). Presses de l'Enssib, Villeurbanne.

Aillerie, K. (2011). Pratiques informationnelles informelles des adolescents (14–18 ans) sur le Web. PhD thesis, Université Paris XIII, Université Paris-Nord, France.

Aillerie, K. (2016). Pratiques informationnelles des jeunes : quels enjeux pour quelle "culture numérique" ? *Champs Culturels*, 28, 41–45.

Béguin-Verbrugge, A. (2006). Pourquoi faut-il étudier les pratiques informelles des apprenants en matière d'information et de documentation ? In *Savoirs et histoires : deux colloques internationaux en éducation*, Astolfi, J.-P. and Houssaye, J. (eds). Laboratoire de sciences de l'éducation-CIVIIC, Mont-Saint-Aignan.

Capelle, C. (2018). Les représentations des risques numériques en éducation : construction de normes dans les discours en circulation. *Revue COSSI*, 5 [Online]. Available at: https://revue-cossi.info/numeros/n-5-2018-processus-normalisation-durabilite-information/729-5-2018-capelle [Accessed 3 January 2021].

Cardon, D. (2008). Le design de la visibilité. *Réseaux*, 6(152), 93–137.

Cardon, D. (2015). *À quoi rêvent les algorithmes*. Le Seuil, Paris.

Chante, A. (2010). La culture de l'information, un domaine de débats conceptuels. *Les enjeux de l'information et de la communication*, 1, 33–44.

Chaudiron, S. and Ihadjadene, M. (2011). De la recherche de l'information aux pratiques informationnelles. *Études de communication*, 2(35), 13–30.

Cordier, A. (2015). *Grandir connectés : les adolescents et la recherche d'information*. C&F, Caen.

Coutant, A. and Stenger. T. (2011). Introduction. *Hermès La Revue*. 1(59), 9–17.

Dauphin, F. (2012). Culture et pratiques numériques juvéniles : quels usages pour quelles compétences ? *Questions vives*, 7(17), 37–52 [Online]. Available at: https://doi.org/10.4000/questionsvives.988 [Accessed 3 January 2021].

Doueihi, M. (2011). *Pour un humanisme numérique*. Le Seuil, Paris.

Ellison, N.B. and Boyd, D. (2013). Sociality through social network sites. In *The Oxford Handbook of Internet Studies*, Dutton, W.H. (ed.). Oxford University Press, Oxford [Online]. Available at: http://www.academia.edu/7731305/Ellison_N._B._and_boyd_d._2013_._Sociality_through_Social_Network_Sites._In_Dutton_W._H._Ed._The_Oxford_Handbook_of_Internet_Studies._Oxford_Oxford_University_Press_pp._151–172.

Juanals, B. (2003). *La culture de l'information : du livre au numérique*. Hermes Science Lavoisier, Paris.

Le Deuff, O. (2010). Bouillon de cultures : la culture de l'information est-elle un concept international ? In *L'éducation à la culture informationnelle*, Chapron, F. and Delamotte, E. (eds). Presses de l'Enssib, Villeurbanne.

Le Deuff, O. (2011). Éducation et réseaux socionumériques : des environnements qui nécessitent une formation. *Hermès La Revue*, 1(59), 67–73.

Le Deuff, O. (2012). Littératies informationnelles, médiatiques et numériques : de la concurrence à la convergence ? *Études de communication*, 38, 131–147.

Liquète, V. (ed.) (2010). *Médiations*. CNRS Éditions, Paris.

Liquète, V. (2012). *Pratiques informelles et non-formelles d'information des jeunes : comment les considérer dans les structures documentaires et bibliothéconomiques ? Actes de la journée du 10 novembre 2011*. Université de Bordeaux, Bordeaux.

Merzeau, L. (2013). Traces numériques et recrutement : du symptôme au cheminement. In *Traces numériques : de la production à l'interprétation*, Galinon-Mélénec, B. and Zlitni, S. (eds). CNRS Éditions, Paris.

Ministère de l'Éducation Nationale, Robine, F. (2017). Circulaire de mission des documentalistes n° 2017-051 du 28-3-2017 Réseau Canopé [Online]. Available at: https://www.reseau-canope.fr/savoirscdi/metier/textes-reglementaires-pour-exercerle-metier-de-professeur-documentaliste/acces-chronologique-aux-textes-reglementaires/2010-2019/circulaire-n-2017-051-du-28-3-2017.html [Accessed 3 January 2021].

Morin, E. (1990). *Introduction à la pensée complexe*. ESF, Paris.

Pasquier, D. (2005). *Cultures lycéennes : la tyrannie de la majorité*. Éditions Autrement, Paris.

Serres, A. (2010). Éducations aux médias, à l'information et aux TIC : ce qui nous unit est ce qui nous sépare. In *L'éducation à la culture informationnelle*, Chapron, F. and Delamotte, E. (eds). Presses de l'Enssib, Villeurbanne.

Serres, A. (2012). *Dans le labyrinthe : évaluer l'information sur Internet*. C&F, Caen.

Serres, A. (2017). Affiliation intellectuelle et culture numérique : la question du modèle. In *Les affiliations par et avec le numérique*, Liquète, V. and Soumagnac, K. (eds). Hermann, Paris.

Soumagnac, K. and Capelle, C. (2017). Formes d'affiliation dans les pratiques informationnelles des lycéens avec le numériques : le cas des Travaux Personnels Encadrés (TPE). In *Les affiliations par et avec le numérique*, Liquète, V. and Soumagnac, K. (eds). Hermann, Paris.

Watier, P. (2002). *Une introduction à la sociologie compréhensive*. Circé, Belval.

7

MIL as a Tool for Teachers to Prevent Risk and Transmit Digital Culture

7.1. Studying digital technology in schools from the perspective of teachers' representations

Children are exposed to screens at increasingly younger ages as the number of household devices, such as tablets and laptops, continues to increase. Despite the recommendations of child psychiatrist Serge Tisseron (2018), who advocates a ban on screens before the age of three years and the measured use of digital technologies from the first grade, in line with the recommendations of the *Académie des sciences* (Tisseron 2013), we note that some children now have strong digital skills before entering middle school, or even from kindergarten, be it discussion in forums dedicated to the game *Fortnite*, children youtubers, watching videos suggested by a recommendation algorithm or children's choreography on Tik Tok. Cordier's work (2015) on adolescents sheds light on the complexity of their uses and their specific strategies for finding information, while pointing out that students do not feel they are digital experts even though they use these technologies daily. Parents and teachers are aware of the digital risks, but they do not master the mechanics of these often-solitary uses: access to pornography and violent images, child pornography networks, cyber harassment, health risks and so on. The digital world is seen as an uncertain and dangerous space where worrying practices reign, such as the mass digital surveillance noted by Casilli (2010), which is a threat to democracy in

---

Chapter written by Julie PASCAU.

For a color version of all the figures in this chapter, see: www.iste.co.uk/capelle/digitalrisks.zip.

the name of internal security. This idea is taken up in numerous cultural works, such as successful dystopian series like *Black Mirror*, which script the excesses of a society governed by technology which favors narcissistic individual development (Haddouk 2017). However, it is also a cultural movement that has its own codes and promotes creativity. Henry Jenkins has spoken of convergence culture as early as 2006, when he analyzed the coexistence of several complementary media. Fluckiger (2008) defines digital culture as "*the expression of values, knowledge, and practices that involve the use of computerized tools, including practices of media and cultural consumption, communication, and self-expression*".

Both a poison and a remedy, digital technology worries and excites in equal measure. This contradiction has led the French Ministry of Education to address the issue of digital technology from an educational and cultural point of view, and not merely from the point of view of teaching technologies. Schools are firmly turning a digital corner by introducing programs that link digital skills and information culture to basic learning. The importance of media literacy in ministers' speeches has been analyzed by Corroy and Froissart (2018), who note that the MIL is presented "*primarily as a tool for critical analysis of the media*", but that the encouraging speeches "*struggle to be embodied in the programs*". Indeed, this is lamented by Liquète (2018), who believes that the specific disciplinary division, temporalities and spatialities of the French school hinder the introduction of information cultures in the classroom that push students to a certain autonomy through the teaching of uncertainty (Cordier 2015). Moreover, the confusion between media and information literacy (MIL) and digital literacy raises questions (Schneider *et al.* 2015) because MIL has a much broader purpose than just the use of digital tools in the classroom. Faced with this distortion of discourses, as Capelle and Rouissi (2018) point out, teachers:

> Are caught between paradoxical demands: on the one hand, they are subject to a societal demand persuading them to protect themselves and their students from these spaces that threaten their privacy; at the same time, their mission is to educate young people about the media and information within the spaces that young people frequent or may be led to frequent.

Teachers therefore integrate digital technology into the classroom, beginning in elementary school, to support the uses of technology by children who become screen users from a very young age. However, it is interesting to see how primary school teachers view the MIL system from kindergarten onwards and what they think of the introduction of digital technology into the school.

7.1.1. *Why be interested in representations?*

In the framework of a piece of doctoral research, the focus falls on the social representations of digital technology in media and information literacy (MIL), insofar as they are held by the Institution, and on an analysis of the representations of primary school teachers about these two objects. The objective is to understand how teachers represent the link between digital technology and MIL through the analysis of their personal and professional practices and their discourses on these practices. This research starts from the observation that there has been no real change in pedagogical practices with digital technology, despite the numerous institutional texts encouraging teachers to introduce it to the classroom (Fourgous report in 2011, Jules Ferry 3.0 report in 2014, middle school reform of 2015, MIL reference document at the start of the 2016 school year, high school reform programs, introduction of the PIX platform and so on). We also note that classroom analysis of traditional mass media (press, radio and television) is almost absent compared to digital media, in comparison to pedagogical practices in the 1980s, even though students still use them on a daily basis (Jehel 2011). Yet media and information literacy has two components: analysis (media and information literacy) and practice (media literacy) (Jacquinot-Delaunay 2011).

According to Erving Goffman, "*the term 'representation' [...] (is) the totality of an actor's activity, which takes place in a period of time characterized by the actor's continuous presence in front of a given set of observers influenced by this activity*". Analyzing teachers' representations seems relevant when analyzing a pedagogical device because it allows us to address the question of the teacher's role (Goffman 1973a) through a communication apparatus, that is, the act of teaching. The teacher's own representations are therefore useful in understanding the specific relationship that they have with the institutional framework, and also how they envisage the transmission of knowledge with regard to their own uses and practices.

To analyze these representations in all their complexity, several methodologies have been used, stemming from different but complementary theoretical currents, of which discourse analysis will be the common thread (Maingueneau 2002). Three stages of data analysis were carried out, as shown in the following diagram.

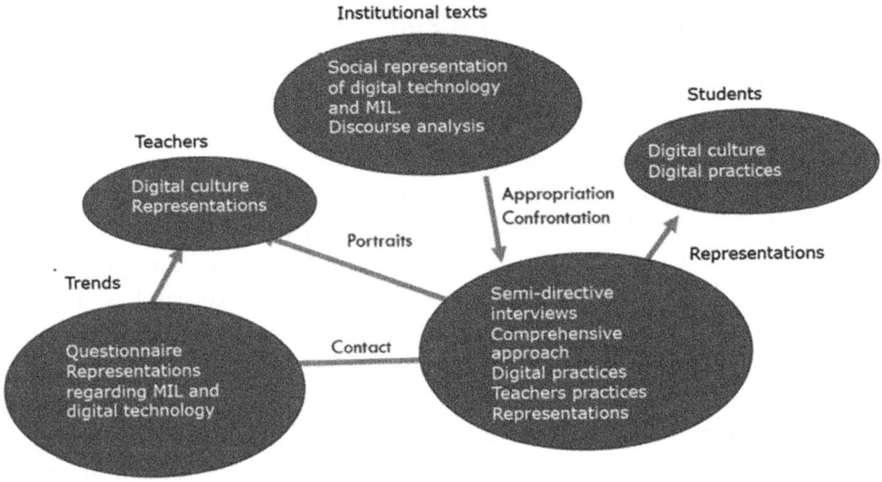

Figure 7.1. *Methodology of research on the representations of digital technology in the context of MIL among primary school teachers*

The first step consists of presenting a framework relating to MIL and digital technology in the institutional discourse since 1980 through the constitution of a corpus of texts, which allows a historical reflection on the evolution of the discourse. The second step consists of identifying the representations of primary school teachers concerning these two objects by means of a questionnaire. Finally, in the last stage, semi-directive interviews are conducted in order to confront the teachers' representations with the framework of institutional discourse and identify their conception of digital culture through the analysis of their uses and practices.

The analysis of the results first focuses on the teachers' relationship with the framework from a semio-pragmatic perspective, in order to understand the reception of institutional discourse among teachers concerning the place of digital technology in schools and the MIL system in particular (Odin 2000). Then, analysis of data from the questionnaire and interviews is performed in two steps: first, to compare their uses, their practices and the representations of MIL that emerge from them by mobilizing the theoretical tools of symbolic interactionism (Goffman 1973; Winkin 1988), and subsequently, to address the question of the construction of the MIL object among the teachers by means of socio-constructivist theories inspired by Jean Piaget and Lev Vygotsky, in order to draw up a schema of what the MIL apparatus encompasses for the teachers.

We can thus propose the initial results from the analysis of the corpus and the questionnaire, as well as a short synthesis of the interview analysis.

7.1.2. *The social representation of digital risks through the analysis of institutional discourses*

The first stage of the work enables us to see the evolution and understand the discourses maintained by the Institution and the international actors around these two nebulas that are digital and media and information literacy. The objective is not to establish an exhaustive and complete analysis of the texts relating to digital and media and information literacy, but to identify the major trends that emerge from them by implementing discourse analysis techniques. To do this, a corpus of 32 texts dated from 1982 to 2017 was established, composed of various types of documents: declarations, laws, plans, programs and so on.

The choice of texts focuses on the important stages in terms of reflection or implementation of digital technology and media literacy in France. A selection of texts in the corpus was compiled to find a quick chronology of the stages of the construction of media literacy in France and to understand the introduction of digital technology from the 1980s to the present day, passing through the stages of computing and new technologies. Most of the texts are in French, but some important international texts, such as the declarations to Unesco, have been selected to underline the link between French advances and European and international reflections that may have had an influence on the normative and discursive construction in France.

The first results show that digital risks occupy a central position in institutional discourses. In 2015, MIL was integrated into French programs, with a strong focus on citizenship through the skills thus put at the service of the construction of a responsible citizen, rather than with the objective of building a "critical understanding" at the service of specific communication, as in the 1980s. Most of the pedagogical materials linked to MIL are oriented towards risk prevention and very few are oriented towards the objective transmission of a digital and media culture. Analysis of the texts shows the shift from media literacy to (digital) MIL, in which the digital realm now involves the term "media" which is eclipsed by the term "information". Thus, the vision of MIL today is different from the vision shown in texts from the 1980s. For example, in 1982, the objective of media literacy, as defined in the Grünwald declaration, was to "*promote among citizens, a critical understanding of the phenomena of communication*"; the terms that stand out in this text are media (18), education (13) and communication (10). At that time, the texts concerning computer science were very separate from anything that had to do with media education. It was not until the arrival of the Internet that media literacy and digital literacy converged. In the year 2000, Le Journal Officiel No. 42 of 23 November, published an insert concerning the "*Brevet informatique et internet école-collège*". This text was addressed to National Education executives to explain the implementation of a new certificate in primary and secondary schools, and to establish the importance of information and communication technologies (ICT) in the new information society. The key words present in this text are mainly focused on ICT, but we can also note the appearance of the keyword "information" with 17 occurrences, at the same level as "computer science", and just behind "Internet" (21) and "brevet" (18). From the 2015 programs and other texts that followed, the term "information" is alternately associated with "media", and the term "digital" with the term "skills". The notion of risks, even though it is not directly present in the texts, is present because of the particular context relating to the revaluation of MIL after the Charlie Hebdo attacks. Indeed, MIL was integrated into a shared foundation and strongly linked to citizenship to prevent the potential risks that may arise when students are left alone with uncritical information.

7.2. What do digital and media literacy evoke in teachers?

In order to begin to gather information from the field and construct the interviews in greater detail, a questionnaire was sent to schoolteachers working in the schools of the Pyrénées Atlantiques department. The first objective was qualitative in order to collect their first representations in relation to digital technology and media literacy, as well as elements of identity in order to interpret the results according to age, gender, training and experience in the teaching profession. Concerning representations relating to digital technology and MIL in schools, the teachers were asked three fairly generic questions: to define media literacy using five key words, to define digital technologies in schools using three key words and to explain the position occupied by digital technologies in the classroom in their professional practice. The analysis of the questionnaire intended for primary school teachers in the Pyrénées Atlantiques department (64) reveals the position held by digital risks in their representations of MIL. The second objective was to identify the teachers who had some sort of digital practice in the classroom, with a view to contacting those who had agreed to be interviewed about their professional background beforehand.

7.2.1. *The weak presence of digital technology and MIL in elementary school*

The position of digital in the classroom in the Pyrénées Atlantiques department is still low, as 40% of respondents never or very rarely use it, compared to 35% nationally in the Profetic 2018 survey[1].

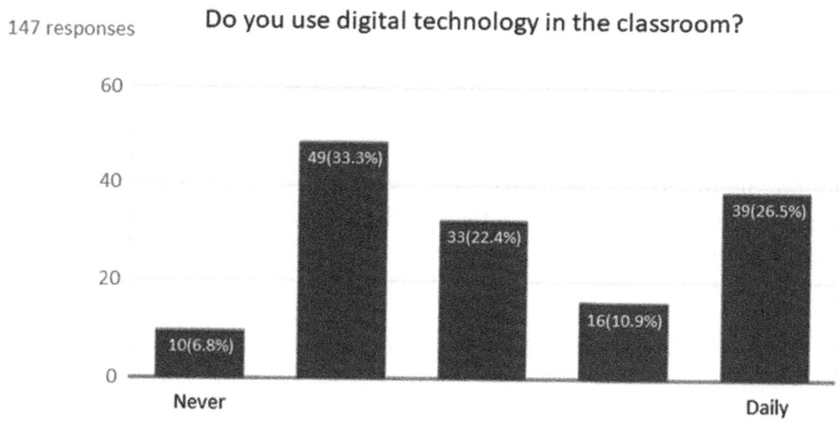

Figure 7.2. *Frequency of digital technology use in the classroom*

1 Ministère de l'Éducation nationale, de la Jeunesse et des Sports – Direction générale de l'enseignement scolaire (2019). L'enquête PROFETIC sur les pratiques numériques des enseignants. *Éduscol* [Online]. Available at: https://eduscol.education.fr/cid60867/l-enquete-profetic.html [Accessed 3 January 2021].

When we try to identify sessions linked to media literacy, we realize that the proportion is very low: 42% of teachers have never conducted a session linked to MIL and among those who have conducted sessions linked to MIL, 34% of respondents only do so very rarely. It is interesting to note that 16% of teachers conduct such sessions regularly and 7.5% of teachers report conducting them very regularly or daily. However, this does not necessarily mean that they do not cover media literacy; it can also reveal that they do not know what media literacy involves in terms of knowledge and skills, and this will therefore be explored in greater depth during the interviews; conversely, it will be relevant to explore what they consider to be media and information literacy.

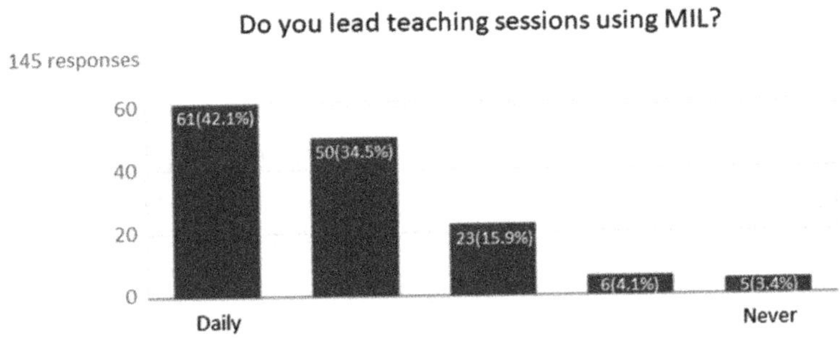

Figure 7.3. *MIL in elementary school, department 64*

Digital technology and MIL still occupy a very limited place in elementary school teaching, even though there is some awareness of the uses and risks for children.

7.2.2. *Risks in the representations of MIL among primary school teachers*

One question deals more specifically with the representations linked to MIL by way of a definition. Primary school teachers were asked to define media literacy using a maximum of five key words. This makes it possible to foresee which words are linked with this subject that they must carry out projects on. The analysis of the initial results shows that the lexical field of digital risks is very present and is one of the major concerns. An initial lexical grouping was created and the first four terms that appear are as follows: critical thinking (32), and – equally – risks (17), prevention (17) (the two terms are closely linked) and citizenship (17). This reveals an initial trend that strongly assimilates media literacy to the idea of a tool for preventing the risks linked to the use of the Internet by young people. The grouping of terms was then done more with the aid of lexicological tools to help build an ontology and make more meaningful groupings. The choice to use the Motbis thesaurus in particular, developed by the Institution of National Education, correlates the framework used by this institution with the proposals of key words by the teachers. In a later phase, an ontology will be constructed to understand how this thesaurus, by virtue of the absent terms or the groupings that it proposes, presents a specific vision of media literacy. The intermediate diagram showing what media and information literacy covers for primary school teachers, constructed with the Motbis thesaurus, can be seen in Figure 7.4.

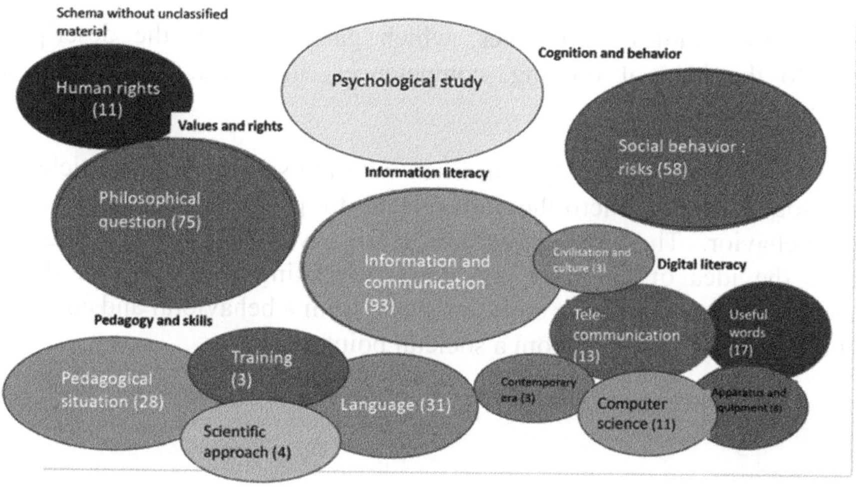

Figure 7.4. *What media and information literacy means to elementary-level teachers of department 64, as described by the Motbis thesaurus (127 responses, translated into English)*

The key words from the questionnaire can be grouped into five main sections, organized first by grouping descriptors in relation to their generic term, and also by grouping micro-thesauri whose ideas are similar. For elementary school teachers, media literacy covers five main areas:

– **Information culture**, which groups together a single micro-thesaurus, that of information and communication. More specifically, it includes elements related to information literacy, media literacy and concepts from the information and communication sciences.

– **Values and rights**, which brings together philosophical issues and human rights. In this grouping, we find values (tolerance and critical reasoning), and also the legislative framework, which notably frames the freedom of expression.

– **Cognition/behavior**, which includes the psychological study and social behavior often related to the risks and dangers linked to the use of digital technology.

– **Digital culture**, which includes everything related to the uses and practices of digital technology in the 21st century.

– **Pedagogy and competences**, which gathers all of the descriptors related to the idea of training, competences, information didactics and pedagogical devices.

Part C is more specifically concerned with the subject of this article and brings together two micro-thesauri related to psychological studies and social behavior. These two micro-thesauri have been brought together because the idea of the risks and dangers relating to the use of digital technology is very present in both sets – one from a behavioral and cognitive point of view and the other from a societal point of view.

Occurrence	Motbis descriptor: specific term (ST)	Microthes auras (MT)	Motbis descriptor: generic term (GT)
53	Cyberviolence (41), Digital divide (5), Social networks (10), society (1)	2310	Social behavior – social life
33	Attention: psychology (6), Curiosity (2), Personality development (1), Imagination (1), Psychological influence (2), Motivation (6), Cognitive process (13), Behavioral psychology (1), Feelings (1)	2205	Psychological study - Psychology GT

Figure 7.5. *Key words to define media literacy grouped by descriptors and micro thesauri from Motbis, and translated into English*

Risks, in the broad sense of the word, were mentioned 58 times in all of the key words, that is, 15% of all of the descriptors selected, in connection with the major idea that stands out, i.e. cyber-violence, which appears 41 times alone, that is, 70% of the risks. The risks are also present in the field of psychology with concerns about attention, personality development and behavior. They are found in other areas too, such as the idea of cybersecurity or the protection of personal data.

To conclude, risks are found in all areas, and this reveals that they are one of the major concerns in the construction of the representation of MIL among teachers.

7.2.3. *A positive perception of the role of digital technology in the classroom*

However, when we analyze their representations of the digital object, we see that the teachers have a positive preconception of its use in the classroom. To the question, "I think that digital technology in the classroom is…", each teacher proposed a set of key words to show their feelings about digital technology. The key words were categorized in a table according to their perception. Each individual is given the status of positive, neutral or negative according to the combination of key words they proposed. This makes it possible to compare the proportion of individuals who have a more positive perception of the totality of the classified words, in order to identify a possible distortion between the two results.

In this analysis, it is important to note that when negative words are proposed, they are often proposed by the same person, which shows a negative vision as a whole. 101 people have a rather positive view of digital, 17 have a neutral view and 17 have a negative view. 75% of respondents have a positive view of digital technology in schools, and only 12.5% have a negative view. 12.5% of respondents have a neutral view, meaning that they balanced their answer with a positive term, counterbalanced with a positive term, or only offered neutral terms.

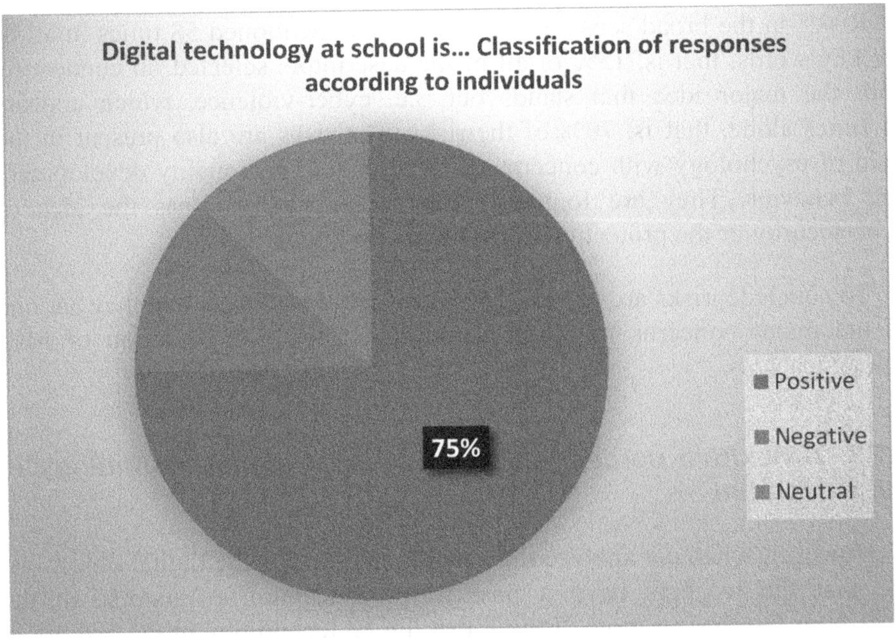

Figure 7.6. *Teachers' feelings about digital technology*

The representation of teachers is partly a reflection of the representation carried by the institution, which remains very oriented towards preventing digital risks and considers digital technology as a tool for the most part, despite the integration of MIL through the skills included in the shared foundation. We can see this in 2018 with the banning of laptops in classrooms: we favor banning and framing over education and reflection when it comes to the use of technology by students. In summary, teachers see MIL more as a tool for information literacy, but the cultural dimension of digital technology and the consideration of young people's digital practices are still absent from these representations.

To the question "I think digital in the classroom is…", three types of key words were proposed: actions, adjectives and key words. There were 120 entries. The first eight terms that stand out very clearly are tools (20), necessary (18), essential (18), useful (16), important (12), unavoidable (9), interesting (10), fun (8), motivating (8) and practical (6). No negative term appears in the first 10 terms. We have to wait until the 11th term to have a negative term appear, difficult (5) and expensive (5). What emerges from this first grouping is that primary school teachers do not seem to be at all reluctant to use digital technology in the classroom; on the contrary, it seems essential to them. However, digital technology is considered above all as a positive pedagogical tool, and not as a culture to be transmitted that can help students think about the world in which they live. The key word tool has been deemed to be neutral in this ranking, because it is ambivalent.

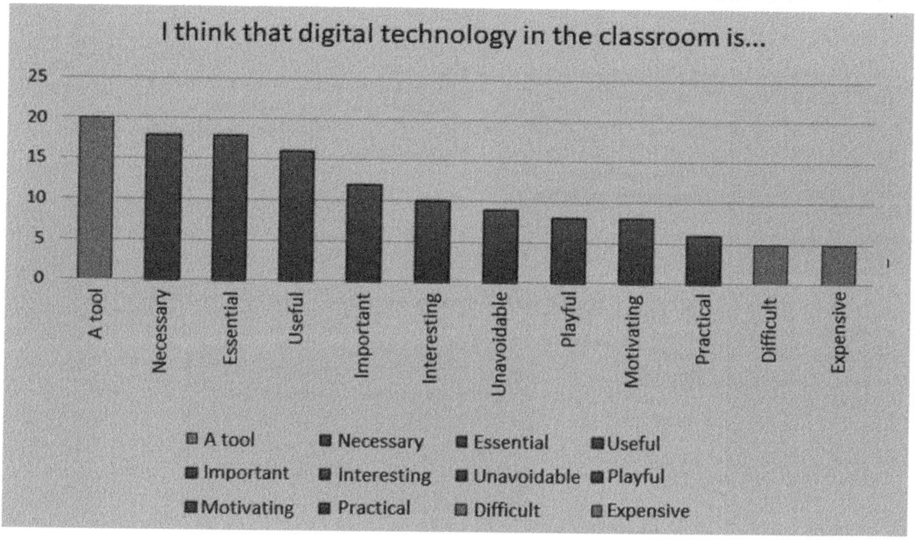

Figure 7.7. *Ranking terms with the most occurrences when the topic of digital technology is raised*

This histogram shows that the negative view trails behind the positive view that teachers have of digital technology.

7.3. The contours of media and information literacy according to teachers

In the context of the interviews, we were able to question the teachers more precisely on what they understand media and information literacy to be and what place digital technology has in this context. Here, we will propose the initial results of the analysis of the teachers' interviews with a greater emphasis on the risks, although these were less apparent in the semi-structured interviews because this was not the main question of the research.

7.3.1. *The objects of MIL from the discourse of primary school teachers*

In the teachers' discourse, when we spontaneously asked them what MIL means for them, they proposed avenues that can be grouped into the five categories shown in the diagram below, along with a synthesis of their personal reflections.

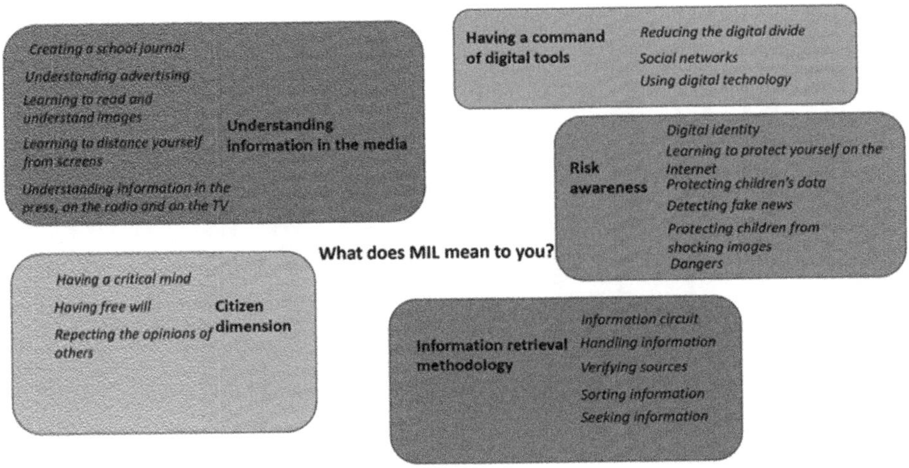

Figure 7.8. *Representation of media and information literacy among elementary school teachers*

Command of digital tools

The use of digital technology in the classroom contributes to MIL by teaching students how to use the tools, and also by teaching the conditions of use for children. The specific issue of social networks can then be addressed in this context. This teaching objective can help reduce the digital divide.

Citizen dimension

Covering MIL in class allows students to develop their critical thinking skills, thus fostering their free will through informed choices, and also, respect for the opinions of others.

Methodology of information retrieval

It is important to train students to search for information, taking care to make them understand the information circuit and familiarize them with all of the steps of checking sources, sorting information and processing it.

Understanding information in the media

Learning to understand information in the media requires knowledge of the different media such as the press, radio and television, mastering the use of screens, and reading and understanding images. This can be done through the creation of a school newspaper or the analysis of commercial media content, such as advertising.

Risk awareness

In MIL, the issue of risks is of great importance to elementary school teachers because the aim is to protect children. The first step is to explain the existing dangers to them; we can teach them how to protect themselves on the Internet and make sure that as teachers, we protect the personal data of the students collected in the classroom. This can be done by making them aware of the issue of digital identity, by informing them of the attitude to adopt when they encounter shocking images such as pornography, but they can also be made aware of the risks inherent in "fake news" and the spreading of rumors.

7.3.2. *What does digital technology mean for teachers?*

In the questionnaire, we asked the same question of elementary school teachers, namely, what digital technology evokes for them, but we asked them to answer in the form of key words. In the interviews, the objective was to develop their discourse by asking them the question without constraining their answers. According to the teachers, digital technology does not necessarily evoke the same aspects, whether they are material or more global. We can distinguish several avenues of what digital technology means according to their personal representation system: the computer tool, the media, dematerialization and access to resources, dangers, and communication and digital culture to be transmitted to students.

Computers as a teaching tool

The most experienced teachers spontaneously respond with the word "computer" or in connection with tools such as the computer, which is less common in the discourse of the youngest teachers. We can see in the speeches that the evolution of the vocabulary that took place in the programs with the word computer, NICT, then lead to another broader dimension, that is, digital tools:

> The word digital for me is anything to do with new technologies, and can mean the use of a computer, PC, tablet, digital camera, and all new technology, and the use that can be made of it in a classroom (Mrs. V.).

Computer equipment and teaching tools are the first ideas that come to the teachers. For most of them, it is a part of their classroom life and they consider it, above all, as a piece of equipment in the classroom, as summarized by Mr. B:

> At school, all that we can do nowadays is thanks to tools that we didn't have in the 90's (Mr. B.).

Dematerialization and access to information

The idea of the dematerialization of resources and access to information is also something that stands out in the discourse of seven teachers. For them, digital is a way to access resources more easily and they see the richness of content that can be accessed in the classroom from a single computer and networks. For Mr. C., who has been a teacher for a long time, it has been completely game changing:

> In a cultural sense, the word dematerialization is interesting and also… I'm not sure how to describe it, but it's something quite incredible when you compare it to the world before. It was something very revolutionary. It's overused as a word, but it's so different and it's not at all the same as the world we lived in before (Mr. C.).

For him, digital technology favors access to culture like never before. Mrs. I. also defines digital from the angle of dematerialization: "*digital technology for me, is a dematerialization of data*". Mrs. IB. also approaches her definition of digital technology from this angle by introducing the idea of information processing:

> It is the encoding of data in a virtual way, which allows the data to be processed in various ways (Ms. IB.).

Mr. M. also takes up the idea of the dematerialization of information and links it to the potential uses that this technology allows:

> Digital technology, I would say that it characterizes a medium, a type of dematerialized medium, mediated by screens and by a material existence. That's what characterizes a medium which we all use for different things: communication, information management, information creation (Mr. M.).

The media

The word media in connection with digital technology is cited only four times and is either linked with the idea of social networks, within the framework of image education with the mention of audiovisuals, or in connection with the idea of media as a teaching aid. Mr. B. has a broader vision, which includes all of the screens with the word multimedia:

> For me, the word digital evokes a whole world that has been open for several decades now, about 20 years for the general public, so everything that concerns screens and multimedia (Mr. B.).

Even though this is not the majority view, digital technology is still linked to the world of media for some teachers who see digital as a media space, rather than as a tool for teaching alone.

Communicate

The idea of communicating is only mentioned by two teachers, like Mrs. AB., who explains that it can open up the class to the outside world "*to communicate with parents, for example, in our digital workspace*" or by Mr. M., who explains that it facilitates communication within the school:

> It has a lot to do with the way teachers communicate with each other. It's the prop that supports school life (Mr. M.).

Dangers

The negative idea of the digital world does not appear to be directly associated with the word digital. Only three teachers mention it: Ms. V. has a positive view of digital technology, but links it directly to the potential abuse that it can generate in children, and makes the link with the obligation to educate children as citizens in this context:

> So a very important aspect, which is becoming more important in my opinion, is that we cannot hide, because the generation of children who are growing up now, they must learn to live with it, along with all of the dangers it can involve, in my opinion. There are misuses that we also need to educate them on, like: this is media literacy, this is how digital technology is used, these are the misuses and dangers. And at my level of kindergarten teaching, I think there are small things that we can put in place right now. For me, it's a type of education that can be done from a young age, with a small downside, which is that digital technology is not suitable for everything. Call me a bit puritanical, but for me it is just a tool. It must remain a tool – a tool with 101 qualities which can have just as many flaws, one of which is the danger element. The danger is definitely there, but that doesn't mean we should conceal digital technology completely. It's essential! (Mrs. V.).

Mr. O., on the contrary, straight away picks up on the idea that digital technology generates a social divide and can create inequalities among students:

> [There's] a social divide! We have a video projector here at the school. But, for example, I was with a pupil this morning who did not want to move because there are children like that and I said: "unfortunately you'll have to come the office to collect a normal keyboard because I don't have a wireless keyboard, so that already puts them at a disadvantage for anything digital. For example, in sixth grade they did their exams on computers: so, there is knowledge, there is the act of finding things on the computer and then there is the use of the keyboard and mouse which causes divides" (Mr. O.).

Mr. B. also reinforces this idea of inequalities in digital culture and uses by explaining that ultimately, everyone uses it without really knowing how it works, which can cause a certain form of slavery in relation to the technology that must be challenged:

> My feeling about digital technology is that everyone has it, everyone is around it, but no one really knows how to use it. It's a tool that we all have, but we don't really have instructions. Whether it's through the internet or screens, I've always said to myself that we are slaves to our digital devices because we don't know how to use them. Essentially, we haven't been given any instructions for how to use them (Mr. B.).

Finally, the relative risks emerging from their representations are strongly linked to the idea that the school has a role to play in media literacy to counter them.

Digital culture to be transmitted to students

The last strong idea that emerges from the digital realm is that there is a culture specific to the digital world to be transmitted. Spontaneously, three teachers explain that it is also important to train students in the specificities of the digital world, as Ms. Pl. does, who only mentions it from the angle of awareness through practice:

> It's very important to make students digitally aware and to make them practice this (Ms. Pl.);

or Ms. I., who goes into more detail:

> It can be the learning of digital technology itself in school as a subject, so that these students gain the theoretical and practical knowledge and learn both the history, the use, how to be alert, what sources to use, etc. (Mrs. I.).

She defends the idea that we can approach it as a subject in its own right. Mr. PG. explains that the use of digital technology in the classroom should be a pretext to address the societal changes that its use brings about, and that this aspect should be taken into account more, especially in the information practices of children, to deepen consideration. He mentions the two aspects of digital education and digital literacy to teach students to maintain a critical distance on their uses.

When teachers are asked what digital technology means to them, the majority see it, above all, as a tool and resource at the service of teaching. But several of them refer to the changes that digital technology in the classroom has brought about in pedagogy, as well as in society, and some of them are convinced that media literacy must be put in place to accompany these uses. Some of them even mention the contours of a digital culture to be transmitted to students. This question was then clearly put to the teachers in order to understand what they understand by this notion.

7.4. What does the requirement to transmit digital culture mean for teachers?

In order to build on the idea of digital culture inscribed in the programs, the teachers were then asked how they understand the assertion of "transmitting digital culture to students", which is recommended by the Ministry of Education as being one of the objectives of MIL. In the framework of the new programs, teachers are indeed asked to transmit a digital culture to students, but this notion is not very explicit, and we thought it was interesting to ask teachers to specify what it evokes for them.

7.4.1. *Digital culture: A very vague concept*

Asking teachers to define what digital culture means in the curriculum confused them. The idea of associating culture with digital seemed strange or awkward to them and, as we can see, the answers were more a search for ideas than straightforward answers. Some of them even admit that they do not understand the expression and do not hesitate to answer that they don't know what it refers to, like Mr. C., who explains that it is a notion that requires reflection and is difficult to explain in simple terms:

> I don't know about that. I can tell you a lot of things, but the thing is that these things must be thought out, they can't just be said during a discussion. In other words, you have to think about them: about their content, their form, etc. (Mr. C.).

Mrs. IB. adds that it is so integrated into our daily lives that we no longer think about it:

> But then it's true that… now, I mean, it's so ingrained that I don't even think about the use of digital technology at its core. We use it; it's become a given, really. (Mrs. IB.).

One of the teachers who spontaneously proposes a precise definition is Mr. M., who wants to build a digital culture that speaks to him and is linked to knowledge:

> It's a culture that confines digital tools to tools, that is, things that enable other things, that are correlated, that meet needs. Trying to rationalize digital for digital's sake. I don't think that an assessment is more interesting if it is done on a tablet than if it is done on paper. There are times when it's better. And then, on the other hand, you can not only develop technical skills, manipulate objects, but also bring knowledge (Mr. M.).

For him, knowing digital technology well means knowing how to evaluate the added value of the tool in the teaching process to serve the student, not just to mobilize digital technology without meaning. To summarize the end of his speech, he proposes distancing the use of these tools through the cultural approach:

> The culture linked to digital technology that I try to transmit, if I intellectualize it a little, is a culture of distancing from social practices linked to digital technology that do not seem to me to be at all reflective, and that seem to me to be dangerous for cultural reasons, because it implies different cultural practices that bring things in some ways and that in other ways are sometimes dangerous. [...] The tools are there, the children use them daily, we use them in class, I think we have to distance ourselves as professionals and think about what we are trying to achieve through all of that. And also get the kids to say, "What are they doing? What does it involve?" (Mr. M.).

This idea of distancing is closely linked to the objectives of media and information literacy.

7.4.2. *What primary school teachers think digital literacy means*

Many ideas were suggested by the teachers, which can be grouped into five more specific categories: information retrieval, dematerialization of data, distancing (critical thinking), the (technological) culture of tools and hardware. The mind map below shows the ideas suggested by the teachers about what they believe represent the cultural aspects of digital technology to be transmitted to the students.

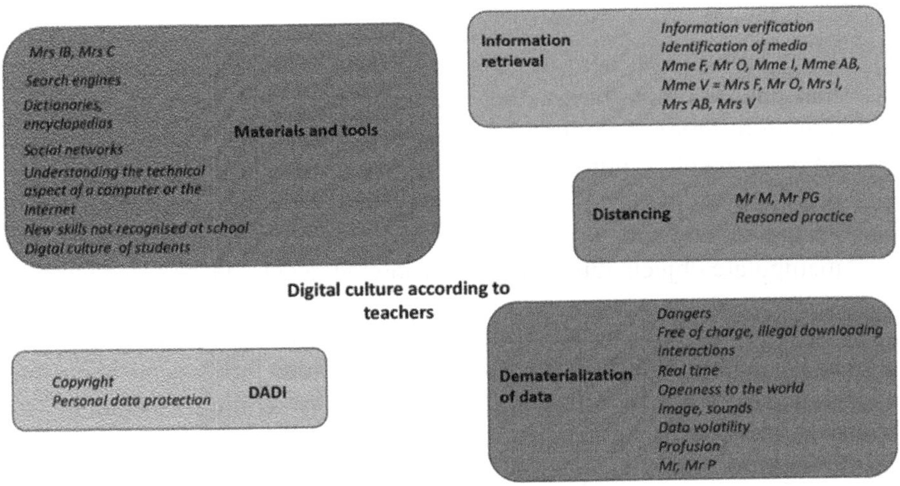

Figure 7.9. *What teachers think digital literacy contains*

The search for information

Learning to search for information is seen as a cultural aspect because it involves certain methodological procedures, from identifying media with a mastery of the notion of source to verifying information, through search operations in the text with the introduction of the notion of key words, or in the search results to be sorted by becoming aware of the multiplicity of content and their irrelevance according to a specific search context. Nearly half of the teachers mention this in connection with the digital culture to be transmitted to students:

> In terms of what we can transmit as a culture, we already know how to search for that: run a search on the computer using key

> words. To think that sometimes when you type on the computer you end up with pornographic pictures, it's a bit embarrassing (Mr. O.).

Indeed, for two of them, training students to search for information is above all a prevention objective, to prevent students from being confronted with pornographic and/or violent content without knowing how to react. The first step in the search for information is to identify the different resources and documents, as well as the forms of memory where knowledge is fixed. The last step in the search for information is to evaluate the resource to gain some critical distance from the information. Mrs. AB. believes that this step must be explained to students because it is essential:

> How to look for information, but also how to process it, to take a step back on whether it's true or not. I think that's already seen as important (Mrs. AB.).

Materials and tools (technical dimensions)

In this section, the teachers question the possible link between the tool and culture. Ms. C. clearly links it to computer equipment, but above all to the ideas that students can develop about the relevance of the uses according to their objectives and representations:

> These are tools: computers, tablets, etc., and maybe the Internet as well. So yes, it's the same thing. What is the culture? What vision do we have of it? Because it also brings us back to that. It's more about the tools and the use we make of them. What tools can I use and when? That's a good idea, maybe we should try to get them to think about it: yes, we can always use the Internet, but maybe it's not always the most useful or relevant, and maybe we should get them to refine their representations or make better use of certain tools, I don't know… (Mrs. C.).

Ms. IB. explains that making students aware of the wealth of tools available can be a relevant way to access this culture:

> It's learning that in the digital world, there are social networks, news media and practical tools for working with images or text. It's learning that it's a tool we can use that in fact accompanies us on a daily basis in fact (Mrs. IB.).

Indeed, as Mr. B. points out, the discretization of the tools makes it difficult to discriminate between all the functions when you only have one terminal to do everything:

> Among young people it is clear that they have everything in one device (Mr. B.).

The vast majority of teachers also believe that digital literacy requires specific knowledge of hardware and tools for accessing information. The first aspect proposed by three male teachers is to explain the technical and material aspect of the functioning of computers, networks and the Internet, so that the students can then understand broader issues:

> At the beginning when I arrived, I had my first computer course, I took apart the computer and I showed them: this is a hard disk drive, I physically showed them the inside of the computer, the machine (Mr. P.).

It is important for these teachers to stop children magically thinking that information is completely dematerialized. They need to understand the tangible side of digital equipment, the fact that, even without wires, it works in relation to an existing but invisible technical system. This then allows us to broach questions related to sustainable development with them in a more concrete way:

> It's true that we all have Internet on our cell phones and nobody knows how it happens. The submarine cables, the consequences of our connections on the planet. I think that this is also part of sustainable development: the servers… We are putting servers in the North Pole or, I don't know, in the north of Norway just so that we can go on Google, Facebook and all of that. So, I think it's important in the digital culture to know that the tools… that we're talking about dematerialization, in fact. It's also important to recognize that it's real material that has material sources and that everything doesn't arrive by magic to our screens (Mr. B.).

Two teachers specifically mention the search engine tool, because, for them, it is a central element that allows access to information, but that one must know and master. Mrs. I. takes up the idea of the search engine, explaining that it is also necessary to reflect with the children on Google's monopoly and to counterbalance it by proposing other search engines in class:

> I think I will bring my own form of digital culture, which would be the notion of search: which search engine? Today there is not just Google; there are several search engines (Mrs. I.).

One teacher also mentions the fact that the knowledge of classic documentary tools that were handled in class can be approached by the students consulting dictionaries or online encyclopedias on the Internet to teach them how to use these tools wisely and to compensate for the lack of equipment with basic publications in certain classes.

The identification of social networks seems important to the teachers because five of them mention them. They are of the view that even though they do not personally use social networks, they should try to understand this tool in order to explain it to the students, so that they know how to discriminate between the tools on the Internet and identify them:

> It's learning that in the digital world, there are social networks, there are news media, and there are also practical tools for image work and text work (Ms. IB.).

At the same time, Mr. B. underlines the fact that social networks convey a culture of their own, which we must learn to decipher, such as the urban legend of Momo who frightened his students, or harassment:

> In particular, there's the Momo thing which is harmful. In terms of what I see on social networks, for middle school students in particular, it's that harassment has taken over on social networks, I think that we forget at a certain point that it's all reality. Just because we're behind a screen doesn't mean that we're not causing harm or affecting people, and I think that the school has a major role to play in improving media literacy around this (Mr. B.).

We must teach them to understand that communication on social networks is based on the same rules of respect as in real life. This is the bridge that Mr. M. also proposes by explaining that digital and social networks require a specific treatment in class, but the learning issues remain the same as in other situations:

> I think that the stakes in terms of learning linked to digital technology overlap with the concerns we may have with regard to other types of media: for example, we talked about the stakes linked to what we post about ourselves on social networks via digital tools, but this issue of education, cooperation and social violence, well these things, you can also find them in the schoolyard, so digital technology requires a different treatment (Mr. M.).

For Mrs. C., children already have a digital culture, and this can be a way to deepen this culture by making them more aware and more critical about their uses:

> Because to transmit a culture, for me they already have that, they already use these tools a lot. Afterwards, if we put culture in the sense of awakened or thoughtful culture behind it, I think it would be rather positive. But it's true that I find that the students are already big users and very fond of technology, and digital technology in general. They already like it a lot more, even the interactive whiteboard, etc., and trivial things too (Mrs. C.).

Mrs. I. takes up this idea very clearly, explaining that for her it is important to take into account the children's representations, in order to work on the basis of their prior knowledge in class:

> I think I would start with the child first and foremost, to understand what their digital literacy is, and when you ask the students a little bit about their digital uses, they know how to identify YouTube and the internet at a very young age (Mrs. I.).

In summary, we can take up Mrs. P.'s ideas. She explains that children develop new operational skills through the handling and use of digital material and tools, and that they should be valued more within the school:

> These are new skills that students are acquiring and that are only very slightly taken into account in their curriculum. If we look from elementary to high school, then it evolves differently because the higher education programs open up to the students and have these options, well, they can opt for these options. But high school students can have other skills, like my generation, for example, in computer science and so on, and it's not for that reason that it's emphasized. So, it's a plus for them. But it can be penalizing too, I think. Because there are children who can be very competent and brilliant in the manipulation of these tools, and it doesn't count for anything. For them, these are skills that are not recognized (Mrs. P.).

Distancing

Three of the youngest teachers in the profession think that transmitting a digital culture has the objective of distancing students from their uses and developing a more reasoned practice. Mr. PG. thinks that:

> Rather than being an unreasonable user of digital technology, transmitting a digital culture, for me, would be to transmit a critical culture, to ensure we start to distance ourselves somewhat from the tools we use (Mr. PG.).

Mrs. C wants "*to encourage them to refine their representations or to make better use of certain tools… and to have a reasoned practice, if not a reasonable one*", while Mr. M. is convinced that:

> We must ensure a certain distancing. The tools are there, the children use them daily, we use them in class, I think we have to distance ourselves as professionals, that is to say, what are we trying to achieve through this? And also have to get the children to say to themselves: what are they doing? What does it involve? What I was saying about materiality, that's part of this distancing.

DADI (Copyright and Information Law)

Ms. I. is the only teacher to mention law as part of the digital culture to be transmitted to students. She specifically mentions two aspects, namely, personal data protection and copyright. For her, this is something to explain to students who have always lived in a space where the majority of resources are dematerialized:

> Talking about music, for example, the notion of copyright also exists on the Internet. I will ask myself questions about digital culture, about their relationship to the notion of private life and public life, and how we can protect this data. What data do we choose to show? To a group? To a community? Or to any audience at all? That's what I think I'd go for if I was teaching digital literacy (Ms. I.).

This comment allows us to address ethical issues related to digital technology with children.

Dematerialization of data

The question of the dematerialization of data is very present when considering the concept of digital culture because it has many consequences for children's systems of representation. Mr. C. has several ideas, such as the increase of possible interactions, the opening up of the world, the profusion of data with rich media like images and sounds, and also raises the problem of the volatility of data and the free access to this profusion via illegal downloads, which is also taken up by Mr. B.:

> Oh well, it goes faster, the information goes faster than in classical culture. As opposed to classical culture, what's different? There is an opening up of the world [...]. There are interactions that there weren't before [...]. They are images, they are words, they are sounds… You get lost in it. You could lose a memory stick containing all of your holiday photos before you got them printed, or you might lose them in a house fire, which is rarer. Whereas a hard disk that blows up is more common now. It is also things like that which are part of a digital culture. It is so abundant [...]. It loses a little bit of its charm because everything is so easy, and this accessibility is so important (Mr. C.).

> In fact, everything is dematerialized, whether it is money, images or information. It's true that if there's a blackout tomorrow, and I'm talking about me personally, that we would essentially lose everything we have! (Mr. B.).

> There is an evasive side to gratuity. In fact, there is a side where nothing has any value anymore; there is that side to it. (Mr. B.).

The dangers related to this dematerialization, such as harassment on social networks, were mentioned by Mr. B. and Mr. M., but were taken up by Mrs. V., who explains that it is necessary to know how to frame the uses to protect the children:

> It is also a case of explaining the dangers that it can involve. Young children are complicated, but sometimes we tell them that sometimes, yes, we can have unwanted screens that appear. How can we remedy this? Well, it's not them who do it, but mom and dad have to be vigilant. You shouldn't necessarily be sat in front of a screen all alone at four/five years old; there is also the issue of limiting the time they spend on these devices, which is important to put in place. And for me, this digital culture involves using the fact that it is a great tool, because while it's very interesting, it needs to be supervised (Mrs. V.).

Mr. PG. summarizes that ultimately, transmitting a digital culture means establishing situations of media and information literacy in the forms mentioned above:

> For me, it's linked to media and information literacy. Rather than being an unreasonable user of digital technology, transmitting a digital culture, for me, means transmitting a critical culture and establishing a distance from the tools we use (Mr. PG.).

7.5. Conclusion

The questionnaire and the interviews provide us with an initial overview of the representations of primary school teachers on digital technology and MIL. Although a few teachers conduct educational sessions with MIL-related objectives, we note that they have a very positive view of the integration of digital technology in the classroom and that the only significant drawbacks that stand out are the lack of equipment and training. However, they mainly see digital technology as a set of tools or teaching aids, with an emphasis on its value in teaching. It is not considered as a culture to be transmitted or a vector of culture at first glance. Students do not use digital technology much in class, and this tool is mostly used in the preparation of sessions or in the form of video-projected content. The situations related to digital technology are thus collective and the student tends to consume content rather than create or manipulate it. In the same way, when teachers were asked to define MIL, the notions of information-documentation were present, but the notions related to digital culture remained subtle. What emerges from these initial results is that risks are very present in the definition of MIL, which shows that it is also seen as a means of risk prevention. This confirms the vision conveyed by French institutional texts that mix digital technology and MIL more along the lines of prevention than notions of digital culture.

However, these initial trends identified are strongly nuanced in the interviews. Indeed, when asked to clarify what they mean by MIL and digital literacy, the teachers' answers as a whole provide a fairly complete picture of what MIL is and go far beyond risk prevention alone, even though in professional practice it is not implemented. When we specifically ask them what they mean by "digital culture", we get answers that focus more on the transmission of an information culture than on a preventive approach. The teachers are well aware that the use of digital technology alone is only effective if there is a relevant link with pedagogy and if the challenges of digital tools are considered, even though many of them feel insufficiently trained and equipped to carry out this task. In conclusion, the perception of digital technology as a tool is very positive. Digital technology is perceived as a facilitator in pedagogy, the necessity of which is accepted and not questioned at all. However, they are wary of pedagogical activities where the student manipulates the tools independently (at least this is what emerges from the first interviews). Teachers are rarely in a position of media literacy; they maintain a position of education through media without necessarily reflecting on these practices. The digital world is not at all considered as a cultural object in their professional practices, whereas in their representations, it is perceived as a rich and difficult universe for students to grasp. Yet, we notice that although they consider it interesting and important, they do not transpose it into their practices, due to lack of time, lack of training or lack of interest. There is also the idea that it is more the culture of the students and that they find it difficult to establish themselves in this learning process, not considering themselves competent or legitimate. However, the notion of digital culture is important to take into account in terms of usage, because it does not only "refer to the idea of the acquisition of knowledge and know-how by users, but more generally refers to *the effect of meaning* produced by technical devices and the uses made of them. In concrete terms, this digital culture is the result of a dual process of acculturation to technology and the technicalization of relationships. It would refer to specific behaviors, representations and values, as well as to a renewed relationship to knowledge. It would also find several forms of expression according to individual conditions and histories (Millerand 1999, p. 379). By only viewing digital technology as a tool or a pedagogical prop, the school misses out on a whole cultural and social dimension that would enable students to critically reflect on the world in which they live.

7.6. References

Capelle, C. and Rouissi, S. (2018). Représentations et stratégies de jeunes enseignants face aux réseaux sociaux numériques. *Les cahiers du numérique*, 14(3), 13–34.

Casilli, A. (2010). *Les liaisons numériques : vers une nouvelle sociabilité ?* Le Seuil, Paris.

Cordier, A. (2015). *Grandir connectés : les adolescents et la recherche d'information*. C&F, Caen.

Corroy, L. and Froissart, P. (2018). L'éducation aux médias dans les discours des ministres de l'Éducation (2005–2017). *Questions de communication*, 34(2), 173–188.

Fluckiger, C. (2008). L'école à l'épreuve de la culture numérique des élèves. *Revue française de pédagogie*, 163, 51–61 [Online]. Available at: https://journals.openedition.org/rfp/978 [Accessed 3 January 2021].

Fourgous, J.-M. (2012). Apprendre autrement à l'ère numérique. Se former, collaborer, innover : un nouveau modèle éducatif pour une égalité des chances. Parliamentary report, Paris [Online]. Available at: https://missionfourgous-tice.fr/missionfourgous2/le-rapport-et-le-livret/ [Accessed 3 January 2021].

Goffman, E. (1973a). *La mise en scène de la vie quotidienne, tome 1 : la présentation de soi*. Éditions de Minuit, Paris.

Goffman, E. (1973b). *La mise en scène de la vie quotidienne, tome 2 : les relations en public*. Éditions de Minuit, Paris.

Goffman, E. (2002). La "distance au rôle" en salle d'opération. *Actes de la recherche en sciences sociales*, 3(143), 80–87.

Haddouk, L. (2017). Black Mirror : le narcissisme à l'ère du numérique. *Le Carnet PSY*, 204(1), 27–29.

Jacquinot-Delaunay, G. (2011). On ne naît pas internaute, on le devient... *Hermès La Revue*, 1(59), 75–76.

Jehel, S. (2011). Parents ou médias, qui éduque les préadolescents ? Enquête sur leurs pratiques TV, jeux vidéo, radio, Internet. *Éducation et société*, 41–58.

Jenkins, H. (2013). *La culture de la convergence : des médias au transmedia*. Armand Colin, Paris.

Liquète, V. (2018). Genèse et essor de la culture de l'information au sein du système éducatif français. Vingt ans après, où en est-on ? *Éducation et sociétés*, 41(1), 151–167.

Maingueneau, D. and Charaudeau P. (eds) (2002). *Dictionnaire d'analyse du discours*. Le Seuil, Paris.

Millerand, F. (1999). L'appropriation du courrier électronique en tant que technologie cognitive chez les enseignants chercheurs universitaires. Vers l'émergence d'une culture numérique ? Communications thesis, Université de Montréal, Montreal.

Ministère de l'éducation nationale, de l'enseignement supérieur et de la recherche. (2015). Socle commun de connaissances, de compétences et de culture, Décret n° 2015–372 du 31-3-2015. *J.O. du 02-04-2015* [Online]. Available from: http://www.education.gouv.fr/pid25535/bulletin_officiel.html?cid_bo=87834 [Accessed 3 January 2021].

Odin, R. (2000). *De la fiction*. De Boeck Université, Brussels.

Pene, S. (2014). Jules Ferry 3.0 : bâtir une école créative et juste dans un monde numérique. *Conseil National du numérique*, Paris [Online]. Available at: https://cnnumerique.fr/wp-content/uploads/2014/10/Rapport_CNNum_Education_oct 14.pdf.

Schneider, E., Serres, A., Stalder A. (2015). L'EMI en partage : essai de cartographie des acteurs. Communication, 10° congrès des Enseignants Documentalistes de l'Éducation nationale, Limoges.

Simmonot, B. (2009). Culture informationnelle, culture numérique : au-delà de l'utilitaire. *Les cahiers du numérique*, 3(5), 25–37.

Tisseron, S. (2013). L'enfant et les écrans : un avis de l'Académie des Sciences. *Le Carnet PSY*, 169(2), 1.

Tisseron, S. (2018). "3-6-9-12", l'état des savoirs. In *3-6-9-12 : apprivoiser les écrans et grandir*, Tisseron, S. (ed.). ERES, Toulouse.

Winkin, Y. (1988). Goffman à Baltasound. Politix. *Revue des sciences sociales du politique*, 3–4, 66–70.

Conclusion

Digital uses have become commonplace over the last few decades and the seductive discourse extolling the potential of digital technology have facilitated its acceptance in society. However, technological advances have never ceased to evolve, with each new development bringing its share of new situations and potentially anxiety-provoking problems. The notion of risk, frequently used, corroborates this feeling. By aiming to solve many problems, digital technologies are also inevitably at the origin of new challenges and uncertain effects that Jacques Ellul pointed out in the 1980s[1]. These effects are never exclusively positive or negative: technologies often encompass both. They indirectly orient uses by authorizing or not certain practices, but they are also a source of inventiveness for users who manage to transform or even divert the uses initially planned, to make other uses, no matter their intention, good or bad. Thus, digital technology is not inherently good or bad. Thinking about information and communication technologies leads us rather to try to understand how individuals appropriate them and to what extent these uses transform their daily life.

By seeking to go beyond the opposition between good and bad uses, this book has sought to compare the notion of digital risk with the discourse, representations and practices of actors in different professional fields. The contributions have allowed for a reflection on the uses of digital technology in schools, in companies, in the media and in everyday life, where cell

---

Conclusion written by Camille CAPELLE.

1 Ellul, J. (2012). *Le bluff technologique*. Pluriel, Paris.

phones have become ubiquitous, by analyzing the way in which representations of risk influence practices. Reading these contributions, we can note the diversity of representations and practices regarding risk in a digital context. For the actors, the notion of digital risk refers to heterogeneous situations and also to emotional responses and action strategies that vary according to representations. To conclude this work, we suggest revisiting the trends that have been highlighted and the questions that remain unanswered.

In the field of education, studies by Camille Capelle, Anne Cordier, Adeline Entraygues and Julie Pascau note the various ways in which teachers appropriate and invest in digital technology. Representations of the digital risks for the teacher or their students often appear to be the starting point of pedagogical practices (see Chapter 6) and can, depending on the case, lead to the renunciation of the use of digital tools in the classrom or, on the contrary, to the implementation of uses in a school context aiming at a reinvestment in the current uses of students, outside of school (see Chapter 1). In Chapter 3, Anne Cordier has shown how significantly the ministerial requirements to introduce digital technology into teaching practices weigh on the daily lives of teachers, especially when they are at the beginning of their careers. She also noted the lack of desire on the part of certain teachers to invest in this field. The notion of digital risk appears to be strongly linked to anxiety, to the fear of doing things wrong or simply of not knowing how to do things. Anne Cordier and Julie Pascau remind us that it is down to the initial and ongoing training of teachers, particularly the National Institutes for Higher Education (INSPE), to accompany these practices by means of an indispensable cultural reflection on digital technology, including the various issues at stake, and not merely teaching the technical use of a few tools in training.

Among the digital risks exposed in the typology (see Chapter 1), the informational risk was studied from the point of view of the search for information in the media by adolescents by Gilles Sahut and Sylvia Francisco in Chapter 2, and for business leaders by Dijana Lekic, Anna Lezon Rivière and Madjid Ihadjadène in Chapter 4. On the Internet, the informational risks of misinformation are numerous. In particular, they are likely to exacerbate mistrust of information on the Internet and to make young people miss out on potentially interesting sources for their learning. In companies, managers fear threats such as the disclosure or theft of data or strategic information. Information risks are often perceived as constraints because they imply behavioral changes for everyone in the company. In addition, in Chapter 5, Freddy Tsopfack Fofack and Bernazi Rengou Abdel show how malicious individuals develop communication skills to scam users.

Following on from Anne Cordier's work[2], the daily life of young people remains to be explored in order to understand the way in which these pedagogical practices effectively lead us to rethink the current uses of digital technology within the social life of these young people. Referring to the texts of the French Ministry of Education, Anne Cordier and Julie Pascau call for an essentially cultural approach to digital technology in professional training In education, as well as in the professional world and private life, information, awareness and training regarding the risks associated with digital practices still appear insufficient and necessary. The practices of mediation and mediatization are still largely to be explored, but a few avenues can be suggested here.

The typology of digital risks could be widely used in education, in the professional world and by politicians to stabilize representations that are still very vague and confused for most users. As a mediation tool, the typology of risks can help to stabilize shared knowledge, raise awareness of the diversity of these risks and enable actors to better identify them in a situation. Knowing how to act in the face of one risk or another, in one situation or another, is a task to be carried out with the users themselves. Depending on the fields of activity, this will be in order to accompany them in their practices and find strategies adapted to their needs, their ways of doing things and their representations of the risk, the latter being an instructive starting point for the analysis of situational practices.

The representations and discourse studied in this book can also be a catalyst, enabling policies to implement institutional measures that are better adapted to the reality on the ground. In particular, it seems important to fill the information gap with users to avoid pitfalls (Chapter 5). The pitfalls of misinformation must therefore be addressed in education, in order to develop an information culture within each citizen (Chapter 2).

2 Cordier, A. (2019). Vers une poiëtique de l'être-au-monde-informationnel. Pour une anthropologie de l'information. HDR, defended at the Université Bordeaux Montaigne, France, December 6.

The Covid-19 health crisis has shown us the essential role of digital uses in society (in work, distribution, education or everyday life). It also shows the capacity of individuals, in case of extreme emergency, to overcome the obstacles linked to the representation of risks in order to act and ensure the continuity of economic, educational and social activities.

Postface

Through a diverse range of narratives, fields and approaches, this book demonstrates that the risk linked to digital information – digital in its broadest sense – is a fundamental question of how individuals relate to their information environment. It involves their knowledge and skills and, beyond that, a form of trust in and to them. Thinking about risk means thinking deeply about the relationship of trust that we commit ourselves to when we engage in our own information practices, from searching, to selecting, to reading. This affects each of us, from the most expert to the novice or beginner in any field of knowledge. Integrating trust into this relationship and into our information-seeking strategies means accepting interdependence and reciprocity, as well as recognizing a certain amount of vulnerability among the actors and even the partners in the relationship. And in that same so-called "trust" relationship, risk-taking, as a negotiated and tolerated act, is part of the process of researching, sorting and evaluating information.

This book demonstrates that risk exists on many levels: on the one hand, it is in the interpersonal relationships that individuals maintain, regardless of their title, their position and sometimes even their "being". Here, risk-taking affects the relationship mainly at an emotional, cognitive and even psychological level. These risks are particularly present and burdensome for young people who are in the process of building their own personalities and awareness. On the other hand, relationships with organizations involve risk-taking that is oriented more towards distrusting, challenging and questioning the institution rather than the individuals that make it up. It seems, having read this book, that some of the risks presented were more concerned with

---

Postface written by Vincent LIQUÈTE.

questioning and challenging the organization than the actors associated with it. Finally, at a broader level, risks can be of an institutional nature and are based on a "formal" social structure, that is, the legal and regulatory framework that governs each society, or the normative context specific to a discipline or a field of knowledge, in the epistemological sense. We then see risk-taking that challenges society in the broadest sense and calls into question the very foundations of the social and political bases of our contemporary societies. Post-truth regimes or conspiracy theories are part of this macro-level, in our view, and profoundly question the relationship of trust that is necessary for dialogue (social, pedagogical, interpersonal, etc.). The supposed foundation of "informational trust", which is necessary for any measured and considered kind of risk-taking, is based on relevance, sharing and acceptance of the wealth and diversity of information. Therefore, working on this informational trust also means accepting and facing risks, while making them emerging objects that must be identified, questioned and deconstructed most of the time. Moreover, the contract that links trust to information must be addressed at all levels: between producers and disseminators, between disseminators and mediators, between mediators and teachers, and between the teacher and student seeking information. We can assume that any breach of this contract would lead to resistance, workarounds or further research into wider pools, structures, organizations or individuals, including the most contestable ones in defiance of the established editorial order. The informational risk involved is major and at a tipping point, which some contributors to this book address in their writing.

Finally, cognitively, it is necessary to have confidence (in the professional, the teacher, the information system or the devices) in order to pursue a search for information and to accept that some certainties will be contradicted. Thus, the information horizon, in the sense of Diane Sonnenwald (1999)[1], is based on a principle of "informational" trust for the actor, combining the processes of legitimization, prescription and appropriation of information. The less informational trust there is, and more broadly in the relationship with the other, the more doubt is created, and the more informational and communicational risk is introduced into the (pedagogical) relationship with others.

1 Sonnenwald, D. (1999). Evolving perspectives of human information behavior: Contexts, situations, social networks and information horizons. In *Exploring the Contexts of Information Behavior: Proceedings of the Second International Conference in Information Needs*, Allen, D.K., Wilson, T.D. (eds). Taylor Graham, London.

Since the advent of Web 2.0, we have gradually moved from mere stated informational risks, brandished like threats, to real risk situations, which are sometimes troubling, traumatizing or even dangerous for students and young people, and by extension for the teacher and indeed any individual. Consequently, it becomes central to adopt a comprehensive research-based and investigative approach to the initial and ongoing training of teachers and students. This applies particularly to systematically questioning ready-made mutual representations based on the actions taken, by teachers towards students and vice versa. Media and information educators, municipal or local authority mediators, knowledge mediators (librarians, etc.) are still too involved in operational approaches to information retrieval, simplifying much more complex information realities, on the principles of rationality and efficiency in the content sought, at a time when the process of deconstruction and comparison of narratives is becoming an absolute necessity. We therefore see that existing pedagogical approaches are limited, still confining digital risks solely to the dangers of screens or media. These approaches can even act *de facto* to prohibit activity (even if accompanied) involving questionable or even reprehensible content that nevertheless deserves to be analyzed, then deconstructed and denounced, rather than discarded without any dialogue or prior analysis. For example, the current international health situation (with respect to Covid-19) forces us to question our relationship to information, its research, its appropriation and the beliefs that stem from it around central issues such as epidemiological statistics, mass vaccination, the ways in which viruses spread, etc. Overexposure to screens and to the media ends up wearying, weighing down and weakening susceptible individuals after the novelty and surprise element wear off. However, analysis of informational and/or digital risks must eventually become central to our teachings, programs and social reflection on information.

Ultimately, combining teaching practice with guiding intentions, through practical use of information environments and devices by young people, will certainly make it possible to share experiences and pitfalls, while also deconstructing myths and digital beliefs that still persist.

List of Authors

Camille CAPELLE
Laboratoire IMS
University of Bordeaux
France

Anne CORDIER
Laboratoire CREM
Équipe PIXEL
University of Lorraine
Nancy
France

Adeline ENTRAYGUES
Laboratoire MICA
Bordeaux-Montaigne University
France

Sylvie FRANCISCO
Inspé
Toulouse
France

Madjid IHADJADENE
Laboratoire Paragraphe
University of Paris 8
France

Dijana LEKIC
Laboratoire Paragraphe
University of Paris 8
France

Anna LEZON-RIVIÈRE
Laboratoire Paragraphe
University of Paris 8
France

Vincent LIQUÈTE
Laboratoire IMS
University of Bordeaux
and
Laboratoire MICA
Bordeaux-Montaigne University
France

Franc MORANDI
Laboratoire IMS
University of Bordeaux
France

Julie PASCAU
Laboratoire MICA
Bordeaux-Montaigne University
France

Abdel Bernazi RENGOU
CFRD-SHSE
University of Yaoundé 1
Cameroon

Gilles SAHUT
Lerass
Jean Jaurès University
Toulouse
France

Freddy TSOPFACK FOFACK
URPHISSA and UCS Center
University of Dschang
Cameroon
and
Institut National de la Recherche Scientifique de Montréal
Canada

Index

C, D

Cameroon, 111–115, 118–122, 124–126, 128, 130
credibility judgment, 28
cybercriminals, 112, 114, 121
digital
  culture, 19, 24, 56, 57, 72, 75, 159, 160, 162, 168, 173, 177–180, 182 –188
  literacy, 3, 8, 11, 14, 17, 18, 20, 22, 24, 160, 164, 178, 180, 182, 184, 186, 188
  natives, 9, 24
  technologies, 3, 4, 7–11, 14, 16, 19, 20–22, 24, 30, 49, 56, 58, 61–77, 83, 84, 94, 95, 99, 159–165, 168–180, 184–188
  transformation, 83, 88
disinformation, 5, 8, 27, 29, 31–33, 37, 41, 45–50

F, G, I

formal, 41, 58, 61, 62, 70, 77, 140, 144
grounded theory, 94
informal, 139, 140, 143, 144, 155
information
  and library didactic, 153, 154
  behavior, 47, 49, 84, 106
  culture, 24, 57, 77, 138, 139, 141, 143, 153–155, 160, 168, 188
  management, 85, 99, 103, 175
  practices, 34, 59, 137–141, 143, 144, 146, 148, 150, 152, 154, 155, 177
  risk management, 84, 87, 91, 94, 96, 97, 100, 105, 106
  risks, 5, 84, 85, 88, 89, 91, 92, 94, 95, 97, 100, 104–106
  security, 89, 91, 92, 94–97, 99–101, 103–106

M, N

media and information literacy, 12, 28, 46–49, 56, 59, 75, 137, 144, 154, 159–163, 165–167, 169–173, 178, 179, 187, 188
middle school, 57, 60, 142, 159, 161, 183
mobile phone, 112, 118–121, 126
  scamming, 121
  users, 119–121
new teachers, 4, 9, 10, 21, 23, 24

P, R

processing
  analytic, 32, 33
  heuristic, 32, 42, 44, 48

professional training, 56, 57, 61, 68, 72, 73, 75–77
representation, 3, 4, 10, 21, 22, 42, 56, 59, 61, 75, 77, 138–140, 144, 145, 148, 150, 154, 159, 161–164, 166, 169, 170, 172, 173, 177, 181, 184–188
risk, 3–9, 11–13, 15, 18, 19, 21, 23–25, 27, 28, 31, 32, 39, 41, 45, 46, 48, 55, 64, 66–68, 71, 74, 76, 81, 83–87, 89, 90, 92, 94, 95, 97, 98, 104, 105, 111, 122, 124, 125, 127, 138, 144–148, 150–154, 159, 164, 166, 168, 169, 171, 173, 177, 188
  digital, 3, 4, 7–11, 14, 17, 18, 21, 23, 24, 39, 41, 46, 55, 62, 64, 74, 83, 86–88, 95, 104, 115, 124, 135, 137–139, 159, 163, 165, 166, 170
  epistemic, 32, 37, 46, 47
  obvious, 112, 128
  perceived, 22, 112
  perceptions, 15, 18
  society, 3, 55

S, T

school, 3, 4, 8–10, 12–24, 34, 35, 40, 45, 49, 55, 57, 58, 60, 62–65, 67, 70, 72, 137, 139, 140–145, 147, 149–151, 153, 154, 159–162, 164–167, 169, 171–177, 180, 183–185, 187, 189
  standard, 153
  teacher, 10, 13, 161, 162, 165–167, 171–173, 180, 187
security policy, 89, 90, 92, 96, 104, 105
social media, 12, 27, 29, 30, 92, 141
teacher librarian, 138, 142, 144, 145, 147, 151, 154
teaching, 9, 17, 22, 23, 28, 46, 55–57, 60, 61, 63, 65, 66, 68, 69, 71–77, 154, 160, 161, 165, 166, 172, 174–176, 178, 179, 186, 187
teenagers, 7, 27, 28, 34, 35, 39, 40, 42, 44–49, 60
telecommunications, 83, 93, 96, 106, 111, 112, 114, 121, 122
top managers, 84, 92–94, 96, 97, 106
typology, 4, 7, 10, 46

Other titles from

iSTE

in

Information Systems, Web and Pervasive Computing

2021

EL ASSAD Safwan, BARBA Dominique
Digital Communications 1: Fundamentals and Techniques
Digital Communications 2: Directed and Practical Work

GAUDIN Thierry, MAUREL Marie-Christine, POMEROL Jean-Charles
Chance, Calculation and Life

LE DEUFF Olivier
Hyperdocumentation (Intellectual Technologies Set – Volume 9)

PÉLISSIER Maud
Cultural Commons in the Digital Ecosystem
(Intellectual Technologies Set – Volume 8)

2020

CLIQUET Gérard, with the collaboration of BARAY Jérôme
Location-Based Marketing: Geomarketing and Geolocation

DE FRÉMINVILLE Marie
Cybersecurity and Decision Makers: Data Security and Digital Trust

GEORGE Éric
Digitalization of Society and Socio-political Issues 2: Digital, Information and Research

HELALI Saida
Systems and Network Infrastructure Integration

LOISEAU Hugo, VENTRE Daniel, ADEN Hartmut
Cybersecurity in Humanities and Social Sciences: A Research Methods Approach (Cybersecurity Set – Volume 1)

SEDKAOUI Soraya, KHELFAOUI Mounia
Sharing Economy and Big Data Analytics

SCHMITT Églantine
Big Data: An Art of Decision Making
(Intellectual Technologies Set – Volume 7)

2019

ALBAN Daniel, EYNAUD Philippe, MALAURENT Julien, RICHET Jean-Loup, VITARI Claudio
Information Systems Management: Governance, Urbanization and Alignment

AUGEY Dominique, with the collaboration of ALCARAZ Marina
Digital Information Ecosystems: Smart Press

BATTON-HUBERT Mireille, DESJARDIN Eric, PINET François
Geographic Data Imperfection 1: From Theory to Applications

BRIQUET-DUHAZÉ Sophie, TURCOTTE Catherine
From Reading-Writing Research to Practice

BROCHARD Luigi, KAMATH Vinod, CORBALAN Julita, HOLLAND Scott, MITTELBACH Walter, OTT Michael
Energy-Efficient Computing and Data Centers

CHAMOUX Jean-Pierre
The Digital Era 2: Political Economy Revisited

COCHARD Gérard-Michel
Introduction to Stochastic Processes and Simulation

DUONG Véronique
SEO Management: Methods and Techniques to Achieve Success

GAUCHEREL Cédric, GOUYON Pierre-Henri, DESSALLES Jean-Louis
Information, The Hidden Side of Life

GEORGE Éric
Digitalization of Society and Socio-political Issues 1: Digital, Communication and Culture

GHLALA Riadh
Analytic SQL in SQL Server 2014/2016

JANIER Mathilde, SAINT-DIZIER Patrick
Argument Mining: Linguistic Foundations

SOURIS Marc
Epidemiology and Geography: Principles, Methods and Tools of Spatial Analysis

TOUNSI Wiem
Cyber-Vigilance and Digital Trust: Cyber Security in the Era of Cloud Computing and IoT

2018

ARDUIN Pierre-Emmanuel
Insider Threats
(Advances in Information Systems Set – Volume 10)

CARMÈS Maryse
Digital Organizations Manufacturing: Scripts, Performativity and Semiopolitics
(Intellectual Technologies Set – Volume 5)

CARRÉ Dominique, VIDAL Geneviève
Hyperconnectivity: Economical, Social and Environmental Challenges
(Computing and Connected Society Set – Volume 3)

CHAMOUX Jean-Pierre
The Digital Era 1: Big Data Stakes

DOUAY Nicolas
Urban Planning in the Digital Age
(Intellectual Technologies Set – Volume 6)

FABRE Renaud, BENSOUSSAN Alain
The Digital Factory for Knowledge: Production and Validation of Scientific Results

GAUDIN Thierry, LACROIX Dominique, MAUREL Marie-Christine, POMEROL Jean-Charles
Life Sciences, Information Sciences

GAYARD Laurent
Darknet: Geopolitics and Uses
(Computing and Connected Society Set – Volume 2)

IAFRATE Fernando
Artificial Intelligence and Big Data: The Birth of a New Intelligence
(Advances in Information Systems Set – Volume 8)

LE DEUFF Olivier
Digital Humanities: History and Development
(Intellectual Technologies Set – Volume 4)

MANDRAN Nadine
Traceable Human Experiment Design Research: Theoretical Model and Practical Guide
(Advances in Information Systems Set – Volume 9)

PIVERT Olivier
NoSQL Data Models: Trends and Challenges

ROCHET Claude
Smart Cities: Reality or Fiction

SALEH Imad, AMMI, Mehdi, SZONIECKY Samuel
Challenges of the Internet of Things: Technology, Use, Ethics
(Digital Tools and Uses Set – Volume 7)

SAUVAGNARGUES Sophie
Decision-making in Crisis Situations: Research and Innovation for Optimal Training

SEDKAOUI Soraya
Data Analytics and Big Data

SZONIECKY Samuel
Ecosystems Knowledge: Modeling and Analysis Method for Information and Communication
(Digital Tools and Uses Set – Volume 6)

2017

BOUHAÏ Nasreddine, SALEH Imad
Internet of Things: Evolutions and Innovations
(Digital Tools and Uses Set – Volume 4)

DUONG Véronique
Baidu SEO: Challenges and Intricacies of Marketing in China

LESAS Anne-Marie, MIRANDA Serge
The Art and Science of NFC Programming
(Intellectual Technologies Set – Volume 3)

LIEM André
Prospective Ergonomics
(Human-Machine Interaction Set – Volume 4)

MARSAULT Xavier
Eco-generative Design for Early Stages of Architecture
(Architecture and Computer Science Set – Volume 1)

REYES-GARCIA Everardo
The Image-Interface: Graphical Supports for Visual Information
(Digital Tools and Uses Set – Volume 3)

REYES-GARCIA Everardo, BOUHAÏ Nasreddine
Designing Interactive Hypermedia Systems
(Digital Tools and Uses Set – Volume 2)

SAÏD Karim, BAHRI KORBI Fadia
Asymmetric Alliances and Information Systems:Issues and Prospects
(Advances in Information Systems Set – Volume 7)

SZONIECKY Samuel, BOUHAÏ Nasreddine
Collective Intelligence and Digital Archives: Towards Knowledge Ecosystems
(Digital Tools and Uses Set – Volume 1)

2016

BEN CHOUIKHA Mona
Organizational Design for Knowledge Management

BERTOLO David
Interactions on Digital Tablets in the Context of 3D Geometry Learning
(Human-Machine Interaction Set – Volume 2)

BOUVARD Patricia, SUZANNE Hervé
Collective Intelligence Development in Business

EL FALLAH SEGHROUCHNI Amal, ISHIKAWA Fuyuki, HÉRAULT Laurent, TOKUDA Hideyuki
Enablers for Smart Cities

FABRE Renaud, in collaboration with MESSERSCHMIDT-MARIET Quentin, HOLVOET Margot
New Challenges for Knowledge

GAUDIELLO Ilaria, ZIBETTI Elisabetta
Learning Robotics, with Robotics, by Robotics
(Human-Machine Interaction Set – Volume 3)

HENROTIN Joseph
The Art of War in the Network Age
(Intellectual Technologies Set – Volume 1)

KITAJIMA Munéo
Memory and Action Selection in Human–Machine Interaction
(Human–Machine Interaction Set – Volume 1)

LAGRAÑA Fernando
E-mail and Behavioral Changes: Uses and Misuses of Electronic Communications

LEIGNEL Jean-Louis, UNGARO Thierry, STAAR Adrien
Digital Transformation
(Advances in Information Systems Set – Volume 6)

NOYER Jean-Max
Transformation of Collective Intelligences
(Intellectual Technologies Set – Volume 2)

VENTRE Daniel
Information Warfare – 2nd edition

VITALIS André
The Uncertain Digital Revolution
(Computing and Connected Society Set – Volume 1)

2015

ARDUIN Pierre-Emmanuel, GRUNDSTEIN Michel, ROSENTHAL-SABROUX Camille
Information and Knowledge System
(Advances in Information Systems Set – Volume 2)

BÉRANGER Jérôme
Medical Information Systems Ethics

BRONNER Gérald
Belief and Misbelief Asymmetry on the Internet

IAFRATE Fernando
From Big Data to Smart Data
(Advances in Information Systems Set – Volume 1)

KRICHEN Saoussen, BEN JOUIDA Sihem
Supply Chain Management and its Applications in Computer Science

NEGRE Elsa
Information and Recommender Systems
(Advances in Information Systems Set – Volume 4)

POMEROL Jean-Charles, EPELBOIN Yves, THOURY Claire
MOOCs

SALLES Maryse
Decision-Making and the Information System
(Advances in Information Systems Set – Volume 3)

SAMARA Tarek
ERP and Information Systems: Integration or Disintegration
(Advances in Information Systems Set – Volume 5)

2014

DINET Jérôme
Information Retrieval in Digital Environments

HÉNO Raphaële, CHANDELIER Laure
3D Modeling of Buildings: Outstanding Sites

KEMBELLEC Gérald, CHARTRON Ghislaine, SALEH Imad
Recommender Systems

MATHIAN Hélène, SANDERS Lena
Spatio-temporal Approaches: Geographic Objects and Change Process

PLANTIN Jean-Christophe
Participatory Mapping

VENTRE Daniel
Chinese Cybersecurity and Defense

2013

BERNIK Igor
Cybercrime and Cyberwarfare

CAPET Philippe, DELAVALLADE Thomas
Information Evaluation

LEBRATY Jean-Fabrice, LOBRE-LEBRATY Katia
Crowdsourcing: One Step Beyond

SALLABERRY Christian
Geographical Information Retrieval in Textual Corpora

2012

BUCHER Bénédicte, LE BER Florence
Innovative Software Development in GIS

GAUSSIER Eric, YVON François
Textual Information Access

STOCKINGER Peter
Audiovisual Archives: Digital Text and Discourse Analysis

VENTRE Daniel
Cyber Conflict

2011

BANOS Arnaud, THÉVENIN Thomas
Geographical Information and Urban Transport Systems

DAUPHINÉ André
Fractal Geography

LEMBERGER Pirmin, MOREL Mederic
Managing Complexity of Information Systems

STOCKINGER Peter
Introduction to Audiovisual Archives

STOCKINGER Peter
Digital Audiovisual Archives

VENTRE Daniel
Cyberwar and Information Warfare

2010

BONNET Pierre
Enterprise Data Governance

BRUNET Roger
Sustainable Geography

CARREGA Pierre
Geographical Information and Climatology

CAUVIN Colette, ESCOBAR Francisco, SERRADJ Aziz
Thematic Cartography – 3-volume series
Thematic Cartography and Transformations – Volume 1
Cartography and the Impact of the Quantitative Revolution – Volume 2
New Approaches in Thematic Cartography – Volume 3

LANGLOIS Patrice
Simulation of Complex Systems in GIS

MATHIS Philippe
Graphs and Networks – 2nd edition

THERIAULT Marius, DES ROSIERS François
Modeling Urban Dynamics

2009

BONNET Pierre, DETAVERNIER Jean-Michel, VAUQUIER Dominique
Sustainable IT Architecture: the Progressive Way of Overhauling Information Systems with SOA

PAPY Fabrice
Information Science

RIVARD François, ABOU HARB Georges, MERET Philippe
The Transverse Information System

ROCHE Stéphane, CARON Claude
Organizational Facets of GIS

2008

BRUGNOT Gérard
Spatial Management of Risks

FINKE Gerd
Operations Research and Networks

GUERMOND Yves
Modeling Process in Geography

KANEVSKI Michael
Advanced Mapping of Environmental Data

MANOUVRIER Bernard, LAURENT Ménard
Application Integration: EAI, B2B, BPM and SOA

PAPY Fabrice
Digital Libraries

2007

DOBESCH Hartwig, DUMOLARD Pierre, DYRAS Izabela
Spatial Interpolation for Climate Data

SANDERS Lena
Models in Spatial Analysis

2006

CLIQUET Gérard
Geomarketing

CORNIOU Jean-Pierre
Looking Back and Going Forward in IT

DEVILLERS Rodolphe, JEANSOULIN Robert
Fundamentals of Spatial Data Quality